Modelling of Wireless Power Transfer

Modelling of Wireless Power Transfer

Editors

Ben Minnaert
Mauro Mongiardo

MDPI • Basel • Beijing • Wuhan • Barcelona • Belgrade • Manchester • Tokyo • Cluj • Tianjin

Editors
Ben Minnaert
Odisee University College
Belgium

Mauro Mongiardo
University of Perugia
Italy

Editorial Office
MDPI
St. Alban-Anlage 66
4052 Basel, Switzerland

This is a reprint of articles from the Special Issue published online in the open access journal *Energies* (ISSN 1996-1073) (available at: https://www.mdpi.com/journal/energies/special_issues/modelling_wireless_power_transfer).

For citation purposes, cite each article independently as indicated on the article page online and as indicated below:

LastName, A.A.; LastName, B.B.; LastName, C.C. Article Title. *Journal Name* **Year**, *Volume Number*, Page Range.

ISBN 978-3-0365-0508-4 (Hbk)
ISBN 978-3-0365-0509-1 (PDF)

© 2021 by the authors. Articles in this book are Open Access and distributed under the Creative Commons Attribution (CC BY) license, which allows users to download, copy and build upon published articles, as long as the author and publisher are properly credited, which ensures maximum dissemination and a wider impact of our publications.

The book as a whole is distributed by MDPI under the terms and conditions of the Creative Commons license CC BY-NC-ND.

Contents

About the Editors vii

Preface to "Modelling of Wireless Power Transfer" ix

Mauro Mongiardo, Giuseppina Monti, Ben Minnaert, Alessandra Costanzo and Luciano Tarricone
Optimal Terminations for a Single-Input Multiple-Output Resonant Inductive WPT Link
Reprinted from: *Energies* **2020**, *13*, 5157, doi:10.3390/en13195157 1

Ben Minnaert, Alessandra Costanzo, Giuseppina Monti, and Mauro Mongiardo
Capacitive Wireless Power Transfer with Multiple Transmitters: Efficiency Optimization
Reprinted from: *Energies* **2020**, *13*, 3482, doi:10.3390/en13133482 23

Aam Muharam, Suziana Ahmad and Reiji Hattori
Scaling-Factor and Design Guidelines for Shielded-Capacitive Power Transfer
Reprinted from: *Energies* **2020**, *13*, 4240, doi:10.3390/en13164240 41

Changping Li, Bo Wang, Ruining Huang and Ying Yi
A Resonant Coupling Power Transfer System Using Two Driving Coils
Reprinted from: *Energies* **2019**, *12*, , doi:10.3390/en12152914 63

Seon-Jae Jeon and Dong-Wook Seo
Coupling Coefficient Measurement Method with Simple Procedures Using a Two-Port Network Analyzer for a Multi-Coil WPT System
Reprinted from: *Energies* **2019**, *12*, , doi:10.3390/en12203950 75

Adam Steckiewicz, Jacek Maciej Stankiewicz and Agnieszka Choroszucho
Numerical and Circuit Modeling of the Low-Power Periodic WPT Systems
Reprinted from: *Energies* **2020**, *13*, , doi:10.3390/en13102651 85

Feng Wen and Rui Li
Parameter Analysis and Optimization of Class-E Power Amplifier Used in Wireless Power Transfer System
Reprinted from: *Energies* **2019**, *12*, 3240, doi:10.3390/en12173240 103

Koen Bastiaens, Dave C. J. Krop, Sultan Jumayev and Elena A. Lomonova
Optimal Design and Comparison of High-Frequency Resonant and Non-Resonant Rotary Transformers
Reprinted from: *Energies* **2020**, *13*, 929, doi:10.3390/en13040929 117

About the Editors

Ben Minnaert obtained his Ph.D. in Engineering in 2007 at Ghent University, Belgium. He has authored or co-authored more than 50 papers on international journals and conferences. In 2018, he obtained a permanent position as a researcher and lecturer at the University College Odisee, KU Leuven Association. His main research interest is the modelling of energy systems, including energy harvesting, photovoltaic solar cells, and wireless power transfer. Recently, he has developed nearfield wireless power transfer systems for nonstatic applications. His research activities are dedicated to embedded systems, wireless sensor networks, IoT applications, and (inductive and capacitive) wireless power transfer for multiple transmitters and receivers.

Mauro Mongiardo (F'11) received the Laurea degree (110/110 cum laude) in Electronic Engineering from the University of Rome "La Sapienza" in 1983. In 1991, he became Associate Professor of Electromagnetic Fields at the University of Perugia; since 2001, he is has been Full Professor of Electromagnetic Fields at the same university. He was elected Fellow of the IEEE "for contributions to the modal analysis of complex electromagnetic structures" in 2011. The scientific interests of Mauro Mongiardo primarily concern the numerical modeling of electromagnetic wave propagation both in closed and in open structures. His research interests involve CAD and optimization of microwave components and antennas. Mauro Mongiardo has served on the Technical Program Committee of the IEEE International Microwave Symposium since 1992. Since 1994, he has is a member of the Editorial Board of the *IEEE Transactions on Microwave Theory and Techniques*. During the years 2008–2010, he was an Associate Editor of the *IEEE Transactions on Microwave Theory and Techniques*. He is an author or co-author of over 200 papers and articles in the fields of microwave components, microwave CAD, and antennas. He is co-author of the books *Open Electromagnetic Waveguides* (IEEE, 1997) and *Electromagnetic Field Computation by Network Methods* (Springer, 2009).

Preface to "Modelling of Wireless Power Transfer"

Wireless power transfer (WPT) allows the transfer of energy from a transmitter to a receiver across an air gap, without any electrical connections. Technically, any device that needs power can become an application for WPT. The current list of applications in which WPT is applied is therefore very diverse, from low-power portable electronics and household devices to high-power industrial automation and electric vehicles. With the rise of IoT sensor networks and Industry 4.0, the presence of WPT will only increase.

In order to improve the current state of the art, models are being developed and tested experimentally. Such models represent either part of the WPT technology or are focused on a certain application. They allow simulating, quantifying, predicting, or visualizing certain aspects of the power transfer from transmitter(s) to receiver(s). Moreover, they often result in a better understanding of the fundamentals of the wireless link.

This book presents a collection of peer-reviewed papers that focus on the modelling of wireless power transmission. It covers both inductive and capacitive wireless coupling and includes work on multiple transmitters and/or receivers. We hope the readers will be able to apply the research results herein to enhance the technology and allow its further implementation into our society.

Finally, we congratulate and thank the authors, reviewers, *Energies* journal, and the MDPI publishers and press production team. This book is a result of their support and efforts.

Ben Minnaert, Mauro Mongiardo
Editors

Article

Optimal Terminations for a Single-Input Multiple-Output Resonant Inductive WPT Link

Giuseppina Monti [1,*,†], Mauro Mongiardo [2,†], Ben Minnaert [3], Alessandra Costanzo [4] and Luciano Tarricone [1]

1. Department of Engineering for Innovation, University of Salento, 73100 Lecce, Italy; luciano.tarricone@unisalento.it
2. Department of Engineering, University of Perugia, 06123 Perugia, Italy; mauro.mongiardo@unipg.it
3. Department of Industrial Science and Technology, Odisee University College of Applied Sciences, 9000 Ghent, Belgium; ben.minnaert@odisee.be
4. Department of Electrical, Electronic and Information Engineering Guglielmo Marconi, University of Bologna, 40126 Bologna, Italy; alessandra.costanzo@unibo.it
* Correspondence: giuseppina.monti@unisalento.it
† These authors contributed equally to this work.

Received: 28 May 2020; Accepted: 22 September 2020; Published: 3 October 2020

Abstract: This paper analyzes a resonant inductive wireless power transfer link using a single transmitter and multiple receivers. The link is described as an $(N+1)$–port network and the problem of efficiency maximization is formulated as a generalized eigenvalue problem. It is shown that the desired solution can be derived through simple algebraic operations on the impedance matrix of the link. The analytical expressions of the loads and the generator impedances that maximize the efficiency are derived and discussed. It is demonstrated that the maximum realizable efficiency of the link does not depend on the coupling among the receivers that can be always compensated. Circuital simulation results validating the presented theory are reported and discussed.

Keywords: resonant; wireless power transfer; inductive coupling; optimal load; single-input multiple-output; power gain

1. Introduction

In recent years, several applications have been proposed for resonant inductive Wireless Power Transfer (WPT) [1–4]. In fact, resonant inductive WPT is an effective solution for wirelessly energizing electronic devices and several optimal design strategies have been investigated in the literature.

Usually, the goal is to recharge a single device and the focus is on maximizing either the power delivered to the load or the power transfer efficiency. In this regard, the most widely adopted scheme is that using a single transmitter, thus corresponding to a Single-Input Single Output (SISO) configuration. In a SISO configuration the link consists of just two magnetically coupled resonators: a transmitting resonator connected to the source and a receiving resonator connected to the load (i.e., the device to be recharged). SISO configurations have been widely investigated in the literature and it has been demonstrated that the link has to be terminated on its conjugate image impedances for maximizing both the power on the loads and the efficiency [5–7].

More recently, schemes using multiple transmitters and/or multiple receivers have been also investigated. The use of Multiple Input Single Output (MISO) schemes could be adopted to obtain an almost constant performance on a given area/volume this being useful if the position of the receiver is affected by small uncertainties (as in the case of embedded devices). In this regard, some interesting results are reported in [8] where it is demonstrated that a two-dimensional region of nearly constant power transfer efficiency can be obtained by using four transmitters. In [9] the use of a linear array of transmitters, activated two at time, is suggested for providing a constant output voltage to a load moving along a linear path. The problem of maximizing the efficiency and the power on the load in MISO schemes has been also analyzed and some interesting results have been reported in [10,11]. In particular, in [10] the solution for maximizing the efficiency has been formulated as a convex optimization problem. In [11] the optimal loads for both the maximum power and the maximum efficiency solutions have been presented for the case of a link using either two–transmitter and a single load or a single transmitter and two–load. In [12], a more abstract approach was used to maximize the efficiency by modeling the MISO-WPT system as a linear circuit whose input-output relationship is expressed in terms of a small number of unknown parameters that can be thought of as transimpedances and gains.

As per schemes using a Single Transmitter and Multiple Receivers (SIMO), they are adopted to recharge multiple devices with a single transmitter [13–28]. In [20] the use of a multiple-output scheme is suggested for the recharge of electric vehicles. The problem of maximizing the power delivered to the loads has been solved in [21], where the expressions of the optimal loads have been derived by using the maximum power transfer theorem for an N–port.

As per the problem of efficiency maximization, in [23] the use of suitable matching networks is suggested. In [24], the specific case of a link using two receivers is analyzed and it is demonstrated that for some specific configurations of the receivers it is convenient to use a non-synchronous scheme with receivers resonating at a frequency different from that of the transmitter. In [22] a SIMO system with constant output voltage and operating at 6.78 MHz is presented. The efficiency of the proposed WPT link is optimized by tuning the input voltage at the transmitter side.

In [25], the loads for maximizing the efficiency have been derived from the expression calculated for the case of a link using one receiver and that using two receivers. However, the analysis is performed assuming that the coupling among the receivers can be neglected, this representing a limitation for real applications. The presence of possible couplings among the receivers has been analyzed in [26,27]. It is demonstrated that for given loads a coupling among the receivers can be compensated by using suitable compensating reactances; however, in these papers it is assumed that the loads are given (i.e., they are not optimized).

Finally, for the problem of efficiency maximization, elegant and comprehensive analysis of all possible configurations (i.e., the SIMO, MISO and MIMO configurations) have been presented in [29,30]. A very elegant and general approach is presented in [29]; where, starting from the impedance or scattering matrix of a multiport the efficiency of a generic MIMO-WPT system is expressed by the Rayleigh quotient. However, the method is not applied on an inductive WPT system and the optimal loads are only expressed as function of the port currents and impedance matrix elements. In [30], the optimal loads are derived from the first-order necessary condition consisting of imposing the zeroing of the first-order partial derivatives of the efficiency with respect to the input and output currents. The optimal solution derived in this way is validated by checking the second order derivatives. The developed analysis is general and overcomes some limitations present in the previous literature. For instance, for the SIMO case a generic number of possibly coupled receivers is considered. Similarly, for the MISO case, the formulas are presented for a generic number of possibly coupled transmitters. However, the analysis developed in [30] is based on the assumption that all the couplings among the transmitters and the receivers are purely inductive; this assumption limits the applicability of the approach to practical applications where the conductivity of the propagation channel is negligibly small.

In this paper, referring to the SIMO configuration, similarly to [29], the problem of finding the optimal loads maximizing the efficiency is formulated as a generalized eigenvalue problem. The presented theory is valid for any strictly passive and reciprocal network in SIMO configuration and is applied in detail for the first time in this paper to the case of a resonant inductive WPT link. The application of the presented theory just requires the knowledge of the impedance matrix of the SIMO network that can be the result of measurements, simulations or theoretical derivation.

The network must not satisfy any particular hypothesis except that of being passive and reciprocal; consequently, the proposed approach is also applicable in the case of non-purely inductive couplings (including the case of a propagation channel with non-negligible values of the conductivity).

The general theory is first presented for a generic $(N+1)$–port network in SIMO configuration and then applied to the specific case of a resonant inductive WPT link; the analytical expressions of the complex loads maximizing the efficiency are derived and discussed. Additionally, the importance of suitably selecting the generator impedance for maximizing the total output power corresponding to the maximum efficiency solution is discussed. The correctness of the derived expressions is validated by the results reported in [30] and by numerical data presented in this paper.

The paper is organized as follows. In Section 2 the problem of efficiency maximization is solved for a generic SIMO $(N+1)$–port network. In Section 3 the derived equations are specialized for the case of an inductive WPT link, the optimal expressions of the loads and the generator impedances are reported. In Section 4 theoretical formulas are validated through circuital and full-wave simulations. Finally, some conclusions are drawn in Section 5.

2. Derivation of the Solution for the General Case

The problem analyzed in this paper is a WPT link using a Single-Input Multiple-Output (SIMO) configuration: a single transmitter is wirelessly connected to N receivers. In this section, the general case is analyzed, no specific assumption is made on the coupling mechanism among the transmitter and the receivers, it is only assumed that the network is passive and reciprocal.

By using a network formalism, the link is modeled as an $(N+1)$–port network $\mathfrak{N}$, see Figure 1, described by its impedance matrix $\mathbf{Z}$. The input port is connected to a sinusoidal source V_G with internal impedance Z_G and the output ports are connected to an N-port load $\mathfrak{N}_L$ with impedance matrix $\mathbf{Z}_L$. Generally, in practical cases, $\mathfrak{N}_L$ consists of a set of N uncoupled load impedances and, consequently, $\mathbf{Z}_L$ is a diagonal matrix

$$\mathbf{Z}_L = \mathrm{diag}(Z_{L,n}), \tag{1}$$

with $n = 1, \ldots, N$.

In real applications, the generator could be a complex network, comprising a DC-AC converter and other circuitry. Accordingly, in general, Z_G is the input impedance of the network adopted for generating the power to be provided at the input port of the network $\mathfrak{N}$. The same consideration applies for each load. In fact, in real applications each load can be a more or less complicated network which in most cases includes a rectifier for converting the AC power at the output port of the network into a DC signal. Accordingly, the generic impedance Z_{Li} is the input impedance of the network connected to the output port i of the link.

The vectors of voltage and current phasors at the network ports, $\mathbf{V}$ and $\mathbf{I}$, and the matrix $\mathbf{Z}$ can be partitioned as

$$\begin{bmatrix} V_i \\ \mathbf{V}_o \end{bmatrix} = \begin{bmatrix} Z_{ii} & \mathbf{Z}_{io} \\ \mathbf{Z}_{oi} & \mathbf{Z}_{oo} \end{bmatrix} \begin{bmatrix} I_i \\ \mathbf{I}_o \end{bmatrix} \tag{2}$$

where V_i and I_i represent voltage and current at the input port, while $\mathbf{V}_o$ and $\mathbf{I}_o$ are the N-vectors of voltages and currents at the output ports.

By replacing the load equation

$$\mathbf{V}_o = -\mathbf{Z}_L \mathbf{I}_o \tag{3}$$

in (2), and by eliminating $\mathbf{I}_o$, the impedance seen at the input port of $\mathfrak{N}$ can be derived as

$$Z_{in} = \frac{V_i}{I_i} = Z_{ii} - \mathbf{Z}_{io}\left(\mathbf{Z}_{oo} + \mathbf{Z}_L\right)^{-1}\mathbf{Z}_{oi}. \tag{4}$$

In a similar way, by combining (2) with the source equation

$$V_i = V_G - Z_G I_i \tag{5}$$

and eliminating I_i, the relation between voltages and currents at the output ports can be cast in the form

$$\mathbf{V}_o = \mathbf{V}_{th} + \mathbf{Z}_{out} \mathbf{I}_o \tag{6}$$

where

$$\mathbf{V}_{th} = \frac{\mathbf{Z}_{oi} V_G}{Z_{ii} + Z_G} \tag{7}$$

is a set of N Thévenin equivalent voltage sources and

$$\mathbf{Z}_{out} = \mathbf{Z}_{oo} - \frac{\mathbf{Z}_{oi} \mathbf{Z}_{io}}{Z_{ii} + Z_G} \tag{8}$$

is the equivalent impedance matrix of the network $\mathfrak{N}$ with the input port closed on the impedance Z_G. The network $\mathfrak{N}$ can be thus represented by the equivalent circuit of Figure 2.

The maximum power transfer between the source and the input port of $\mathfrak{N}$ can be achieved when the conjugate match condition

$$Z_G = Z_{in}^*, \tag{9}$$

where $*$ denotes conjugation, is satisfied. In this case, the power delivered to $\mathfrak{N}$ is equal to the generator available power

$$P_{AG} = \frac{|V_G|^2}{8 \operatorname{Re}[Z_G]}. \tag{10}$$

As far as the output side is concerned, it can be proved [31] that the power delivered by $\mathfrak{N}$ is maximized when the output currents $\mathbf{I}_o$ assume the values $\mathbf{I}_{oM}$ given by

$$\mathbf{I}_{oM} = -\left(\mathbf{Z}_{out} + \mathbf{Z}_{out}^\dagger\right)^{-1} \mathbf{V}_{th} \tag{11}$$

where † denotes conjugate transpose, and consequently the available power at the output ports of $\mathfrak{N}$ is

$$P_a = \frac{1}{4} \mathbf{V}_{th}^\dagger \left(\mathbf{Z}_{out} + \mathbf{Z}_{out}^\dagger\right)^{-1} \mathbf{V}_{th}. \tag{12}$$

It can be noted that for $N > 1$, the optimal load is not univocally defined. In fact, the optimal currents can be obtained by any impedance matrix $\mathbf{Z}_{Lm}$ such that

$$\mathbf{Z}_{LM} \mathbf{I}_{oM} = \mathbf{Z}_{out}^\dagger \mathbf{I}_{oM} = -\mathbf{V}_{oM} \tag{13}$$

where $\mathbf{V}_{oM}$ are the voltages at output ports for $\mathbf{I}_o = \mathbf{I}_{oM}$. Equation (13) also shows that it is possible to realize $\mathbf{Z}_{LM}$ as a set of N independent passive impedances provided that the possible zero elements of $\mathbf{I}_{oM}$ corresponds to zero elements of $\mathbf{V}_{oM}$, and that the phase difference between any two corresponding elements of $\mathbf{I}_{oM}$ and $\mathbf{V}_{oM}$ is $\geq 90°$ in absolute value.

According to the previous discussion, also the problem of determining the impedances Z_G and $\mathbf{Z}_L$ which provide the simultaneous maximum power transfer at the input and output ports has not a unique solution.

To simplify the calculation of the optimal terminations, it is convenient to determine the corresponding optimal currents, which, on the contrary, are univocally defined.

Making use of (2), the total power delivered to the loads P_o, i.e., the sum of the powers delivered to each load P_{oi}

$$P_o = \sum_{n=1}^{N} P_{oi}, \qquad (14)$$

can be expressed as a function of the port currents as

$$\begin{aligned} P_o &= -\frac{1}{4}\left(\mathbf{V}_o^\dagger \mathbf{I}_o + \mathbf{I}_o^\dagger \mathbf{V}_o\right) = \\ &= -\frac{1}{4}\left[\mathbf{I}_o^\dagger\left(\mathbf{Z}_{oo} + \mathbf{Z}_{oo}^\dagger\right)\mathbf{I}_o + \mathbf{I}_o^\dagger \mathbf{Z}_{oi} I_i + I_i^* \mathbf{Z}_{oi}^\dagger \mathbf{I}_o\right] \end{aligned} \qquad (15)$$

and, similarly, the input power can be expressed as

$$\begin{aligned} P_i &= \frac{1}{4}\left(V_i^* I_i + I_i^* V_i\right) = \\ &= \frac{1}{4}\left[I_i^*\left(Z_{ii} + Z_{ii}^*\right)I_i + I_i^* \mathbf{Z}_{io}\mathbf{I}_o + \mathbf{I}_o^\dagger \mathbf{Z}_{io}^\dagger I_i\right]. \end{aligned} \qquad (16)$$

The previous equations can be cast in the form

$$\begin{aligned} P_o &= \frac{1}{4}\mathbf{I}^\dagger \mathbf{A}\mathbf{I} \\ P_i &= \frac{1}{4}\mathbf{I}^\dagger \mathbf{B}\mathbf{I} \end{aligned} \qquad (17)$$

where the matrices $\mathbf{A}$ and $\mathbf{B}$ are defined as

$$\mathbf{A} = -\begin{bmatrix} 0 & \mathbf{Z}_{oi}^\dagger \\ \hline \mathbf{Z}_{oi} & \mathbf{Z}_{oo} + \mathbf{Z}_{oo}^\dagger \end{bmatrix} \qquad (18)$$

$$\mathbf{B} = \begin{bmatrix} Z_{ii} + Z_{ii}^* & \mathbf{Z}_{io} \\ \hline \mathbf{Z}_{io}^\dagger & 0 \end{bmatrix}. \qquad (19)$$

The power gain of $\mathfrak{N}$, defined as the ratio between the output and the input power, can thus be expressed as

$$G_P = \frac{P_o}{P_i} = \frac{\mathbf{I}^\dagger \mathbf{A}\mathbf{I}}{\mathbf{I}^\dagger \mathbf{B}\mathbf{I}}. \qquad (20)$$

In the context of WPT the quantity expressed in (20) is usually referred to as the efficiency of the link, in this paper it will be referred to as G_P in analogy with the terminology adopted in the context of two-port networks.

The power gain is maximized when the maximum power transfer is realized at the output port. Since G_P is a generalized Rayleigh quotient, the maximum of G_P can be determined by solving a generalized eigenvalue problem.

As a matter of fact, using the quotient rule and taking into account the fact that $\mathbf{A}$ and $\mathbf{B}$ are Hermitian matrices, the differential of G_P can be calculated as

$$\delta G_P = 2\frac{(\delta \mathbf{I}^\dagger \mathbf{A}\mathbf{I})(\mathbf{I}^\dagger \mathbf{B}\mathbf{I}) - (\mathbf{I}^\dagger \mathbf{A}\mathbf{I})(\delta \mathbf{I}^\dagger \mathbf{B}\mathbf{I})}{(\mathbf{I}^\dagger \mathbf{B}\mathbf{I})^2}. \qquad (21)$$

Hence, requiring $\delta G_P = 0$ yields

$$\mathbf{A}\mathbf{I} - \frac{\mathbf{I}^\dagger \mathbf{A}\mathbf{I}}{\mathbf{I}^\dagger \mathbf{B}\mathbf{I}}\mathbf{B}\mathbf{I} = 0, \qquad (22)$$

which can be rewritten as

$$\mathbf{Ax} = \lambda \mathbf{Bx} \tag{23}$$

and can be recognized as a generalized eigenvalue problem with $\lambda = G_p$ being the eigenvalue and $\mathbf{x} = \mathbf{I}$ the corresponding eigenvector.

Since, by hypothesis, $\mathfrak{N}$ is passive and the power supply is provided only at the input port, the maximum power gain, G_M, and the corresponding currents (up to an arbitrary factor) can be determined by solving (23) with the constrains

$$\begin{aligned} \lambda &\leq 1 \\ P_i &\geq 0 \\ P_o &\geq 0. \end{aligned} \tag{24}$$

After determining the optimal currents

$$\mathbf{I}_M = \begin{bmatrix} I_{iM} \\ \mathbf{I}_{oM} \end{bmatrix} \tag{25}$$

by (23), the corresponding voltages

$$\mathbf{V}_M = \begin{bmatrix} V_{iM} \\ \mathbf{V}_{oM} \end{bmatrix} \tag{26}$$

can be obtained by (2). Hence the source impedance providing maximum power transfer at the input port can be calculated by letting

$$Z_{GM} = \frac{V_{iM}^*}{I_{iM}^*}, \tag{27}$$

while the maximum power transfer at the output port is obtained with any load $\mathfrak{N}_L$ whose impedance matrix satisfies (13). In particular, if the previously enunciated conditions are satisfied, $\mathfrak{N}_L$ can be realized as a set of uncoupled loads with impedances

$$Z_{LM,n} = -\frac{V_{oM,n}}{I_{oM,n}} \tag{28}$$

with $n = 1, \ldots, N$.

It is worth observing that the theory presented in this section is completely general, it can be applied to any passive SIMO network; moreover, for its application it is sufficient to know the impedance matrix of the network. It is possible to derive the maximum achievable power gain and the optimal loads starting from the impedance matrix, which can be the results of measurements, theoretical calculation or a numerical analysis. Figure 3 summarizes how to apply the proposed approach for the determination of the load impedances maximizing the efficiency of a SIMO Resonant Inductive WPT Link.

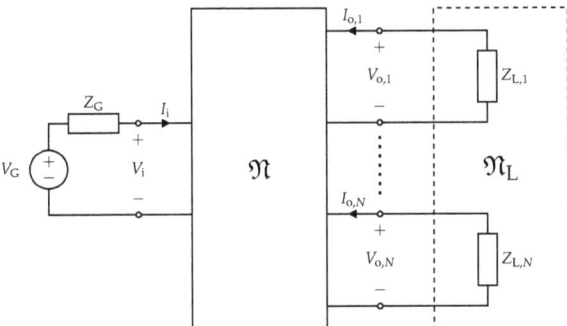

Figure 1. Schematic representation of a SIMO WPT link.

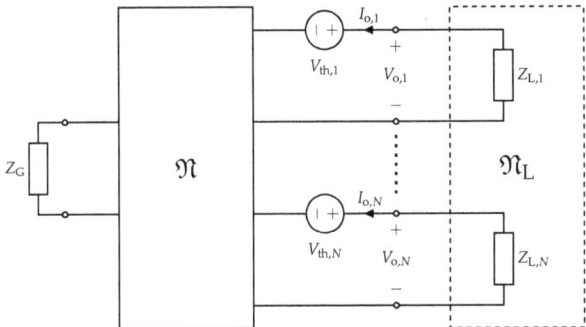

Figure 2. Equivalent Thévenin representation of the circuit of Figure 1.

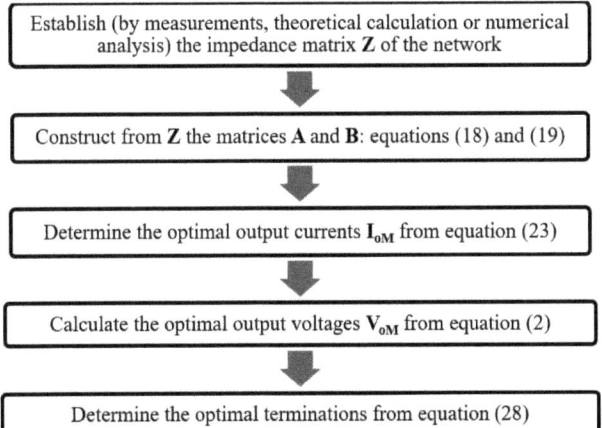

Figure 3. Block diagram of the proposed approach to determine the optimal terminations for efficiency maximization of a SIMO Resonant Inductive WPT Link.

3. The Case of an Inductive Resonant Coupling

In this section, the specific case of a WPT link consisting of $(N+1)$ magnetically coupled resonators is considered (Figure 4). More specifically, the link consists of $(N+1)$ magnetically coupled inductors, L_i, each one loaded by a suitable compensating capacitor, C_i, realizing the resonance condition at the operating angular frequency (i.e., $\omega_0 = 1/\sqrt{L_i C_i}$). The inductor losses are modeled by series resistors R_i related to the quality factors of the coupled resonators:

$$Q_n = \frac{\omega L_n}{R_n}. \tag{29}$$

The coupling between the inductors L_m and L_n is described by the coupling factor k_{mn} related to the mutual inductance M_{mn}

$$k_{mn} = \frac{M_{mn}}{\sqrt{L_m L_n}}. \tag{30}$$

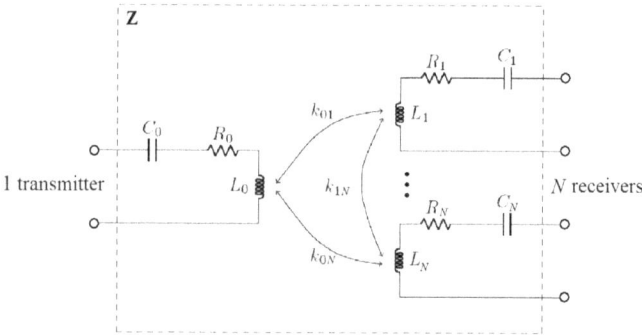

Figure 4. Equivalent circuit of a WPT link with a single transmitter and N receivers, determined by its impedance matrix **Z**.

Accordingly, the network is described by the following impedance matrix:

$$\mathbf{Z} = \begin{bmatrix} R_0 & j\omega M_{01} & j\omega M_{02} & \cdots & j\omega M_{0N} \\ j\omega M_{01} & R_1 & j\omega M_{12} & \cdots & j\omega M_{1N} \\ j\omega M_{02} & j\omega M_{12} & R_2 & \cdots & j\omega M_{2N} \\ \vdots & \vdots & \vdots & \ddots & \vdots \\ j\omega M_{0N} & j\omega M_{1N} & j\omega M_{2N} & \cdots & R_N \end{bmatrix}. \tag{31}$$

By introducing the normalization matrix **d**:

$$\mathbf{d} = \mathrm{diag}\left(\frac{1}{\sqrt{\omega L_n}}\right), n = 0, \ldots, N, \tag{32}$$

it is possible to obtain the following normalized expression for the impedance matrix of the network:

$$\mathbf{z} = \mathbf{dZd} = \begin{bmatrix} \frac{1}{Q_0} & jk_{01} & jk_{02} & \cdots & jk_{0N} \\ jk_{01} & \frac{1}{Q_1} & jk_{12} & \cdots & jk_{1N} \\ jk_{02} & jk_{12} & \frac{1}{Q_2} & \cdots & jk_{2N} \\ \vdots & \vdots & \vdots & \ddots & \vdots \\ jk_{0N} & jk_{1N} & jk_{2N} & \cdots & \frac{1}{Q_N} \end{bmatrix}. \tag{33}$$

Referring to Section 2 and to the Appendix A, for the specific analyzed case it is possible to derive:

$$\begin{aligned}
\tilde{z}_{ii} &= \frac{2}{Q_0} \\
\mathbf{z}_{io} &= \begin{bmatrix} jk_{01} & \cdots & jk_{0N} \end{bmatrix} \\
\mathbf{z}_{oi} &= \mathbf{z}_{io}^T \\
\tilde{\mathbf{z}}_{oo} &= \mathrm{diag}\left(\frac{2}{Q_n}\right)
\end{aligned} \tag{34}$$

and

$$\begin{aligned}
c_0 &= \sum_{n=1}^{N} k_{0n}^2 Q_n \\
c_1 &= -c_0 - \frac{2}{Q_0} \\
c_2 &= c_0.
\end{aligned} \tag{35}$$

Accordingly, by introducing the parameter α:

$$\alpha = \sqrt{1 + \sum_{n=1}^{N} k_{0n}^2 Q_0 Q_n}, \tag{36}$$

for the analyzed case, the solving equation is (see the Appendix A):

$$(\alpha^2 - 1)\lambda^2 - 2(\alpha^2 + 1)\lambda + (\alpha^2 - 1) = 0. \tag{37}$$

Equation (37) has two eigenvalues:

$$G_M = \frac{\alpha - 1}{\alpha + 1}, \quad G_{M1} = \frac{\alpha + 1}{\alpha - 1}. \tag{38}$$

It is evident that $G_{M1} > 1$; as a consequence, only G_M satisfies the first constrain expressed in (24). By choosing to normalize the input current to 1

$$i_{iM} = 1, \tag{39}$$

the following normalized eigenvectors can be obtained

$$i_{oM,n} = -j \frac{k_{0n} Q_n}{\alpha + 1}. \tag{40}$$

The corresponding normalized voltages are:

$$v_{iM} = \frac{\alpha}{Q_0}, \tag{41}$$

$$v_{oM,n} = \frac{1}{\alpha + 1} \left[\left(\sum_{\substack{m=1 \\ m \neq n}}^{N} k_{0m} k_{nm} Q_m \right) + jk_{0n}\alpha \right]. \tag{42}$$

Hence the optimal normalized source impedance is given by:

$$z_{GM} = \frac{v_{iM}^*}{i_{iM}^*} = \frac{\alpha}{Q_0}, \tag{43}$$

however, the optimal N-port load network $\mathfrak{N}_L$ can be realized as a set of uncoupled loads with normalized impedances:

$$z_{\text{LM},n} = -\frac{v_{\text{oM},n}}{i_{\text{oM},n}} = r_{\text{LM},n} - \mathrm{j}\, x_{\text{LM},n} \tag{44}$$

$$r_{\text{LM},n} = \frac{\alpha}{Q_n} \tag{45}$$

$$x_{\text{LM},n} = \frac{1}{k_{0n} Q_n} \sum_{\substack{m=1 \\ m \neq n}}^{N} k_{0m} k_{nm} Q_m \tag{46}$$

The corresponding unnormalized expressions are:

$$R_{\text{LM},n} = \alpha\, R_n \tag{47}$$

$$X_{\text{LM},n} = \frac{R_n}{k_{0n}} \sum_{\substack{m=1 \\ m \neq n}}^{N} k_{0m} k_{nm} Q_m. \tag{48}$$

Discussion of the Results

According to the above reported formulas, the following considerations can be drawn.

- From (38) it is evident that the maximum realizable efficiency of the link only depends on the quality factors of the resonators and on the couplings between the transmitting and the receiving resonators; however, it does not depend on the couplings among the receivers. This means that a possible coupling among the receivers can be always compensated.
- In general, the optimal loads are complex quantities.
- For identical resonators with the same quality factor the real part of the optimal loads is the same for all the loads.
- The imaginary parts of the optimal loads are zero for uncoupled receiving resonators; this means that they play a role of compensation.
- By comparing the expression of the reactive parts of the optimal loads with those of the optimal loads reported in [21] for the maximum power case, it can be easily verified that they are coincident. This means that the same compensating reactances are required for both the maximum efficiency case (i.e., for maximizing the power gain) and the maximum power case.
- The proposed approach also provides the optimal value of the generator impedance, see (27); however, G_p does not depend on the generator. The value provided for Z_G in (27) is that maximizing the power entering the network, and then the power delivered to the loads, when the loads are those maximizing G_p.

It is worth observing that all the achieved results are in a perfect agreement with those reported in [30]. With respect to previously proposed approaches, the theory presented in this paper has the advantage of being completely general, it is valid for any strictly passive and reciprocal network. Additionally, the application of the presented theory just needs the impedance matrix of the link that can be the result of measurements or simulations or theoretical evaluation. In fact, the optimal loads are obtained by solving the eigenvalue problem expressed in (23). To solve the eigenvalue problem one just needs the matrices **A** and **B** that can be directly computed from the **Z** matrix, see (18) and (19).

4. Validation of the Results

To validate the theoretical data, full-wave and circuital simulations have been performed. The commercial tool CST Microwave Studio has been used for full-wave simulations, while the NI AWR Design Environment has been adopted for circuital simulations. Four different WPT links in SIMO configuration have been analyzed. The first three analyzed cases have identical resonators and a different number of receivers, as detailed in the following.

- Case 1: two receivers, see Figure 5a;
- Case 2: three receivers, see Figure 5b;
- Case 3: four receivers, see Figure 5c.

All the analyzed coils (transmitter and receivers) have the same dimensions; they are circular loops with a radius of 5 mm designed by using a copper wire with a radius of 0.3 mm. An operating frequency f_0 of 500 MHz has been assumed. First, the single loop has been analyzed so to calculate the equivalent inductance. From full-wave simulations at f_0 each loop corresponds to an inductance of about 20.9 nH. Accordingly, a series capacitor of 4.84 pF has been added to each loop so to make them resonating at f_0. The relative positions of the transmitting and the receiving coils assumed for the three analyzed cases are illustrated in Figure 5.

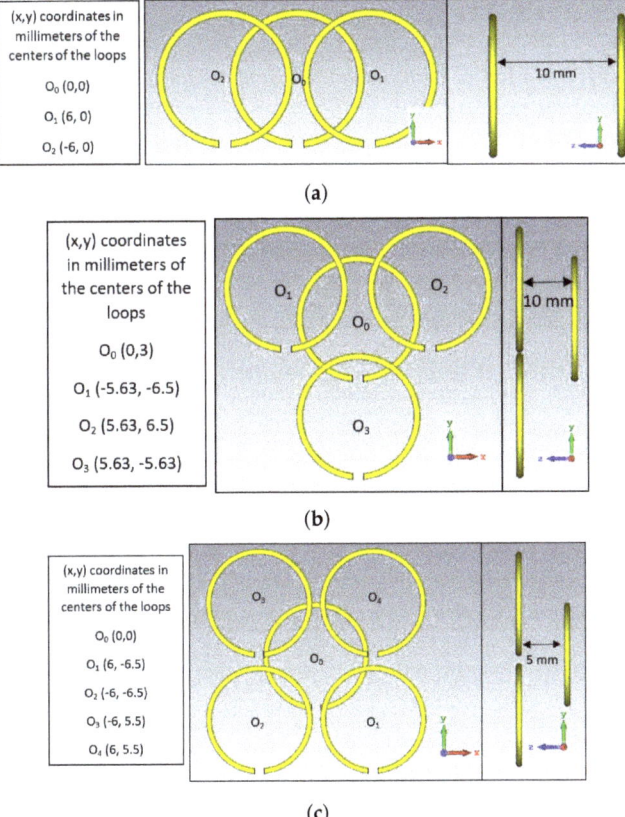

Figure 5. WPT links analyzed through full-wave simulations. (**a**) Case 1: single transmitter and two-receiver link; (**b**) Case 2: single transmitter and three–receiver link; (**c**) Case 3: single transmitter and four-receiver link. In all cases the transmitter is the loop with the center in the point O_0.

To calculate the impedance matrices, each link has been analyzed through full-wave simulations as a multiport network. The following impedance matrices have been obtained:

$$Z_{Case1} = \begin{pmatrix} 0.118 & 1.41\,j & 1.41\,j \\ 1.41\,j & 0.118 & -2.62\,j \\ 1.41\,j & -2.62\,j & 0.118 \end{pmatrix}, \qquad (49)$$

$$Z_{Case2} = \begin{pmatrix} 0.118 & 2.66j & 2.66j & 1.17j \\ 2.66j & 0.118 & -1.84j & -1.86j \\ 2.66j & -1.84j & 0.118 & -1.86j \\ 1.17j & -1.86j & -1.86j & 0.118 \end{pmatrix}, \qquad (50)$$

$$Z_{Case3} = \begin{pmatrix} 0.118 & 0.81j & 0.81j & -1.37j & -1.37j \\ 0.81j & 0.118 & -2.60j & 0.56j & 2.58j \\ 0.81j & -2.60j & 0.118 & 2.58j & 0.56j \\ -1.37j & 0.56j & 2.58j & 0.118 & -2.60j \\ -1.37j & 2.58j & 0.56j & -2.60j & 0.118 \end{pmatrix}. \qquad (51)$$

By comparing the general expression of the impedance matrix of a resonant inductive WPT link given in (31) with the numerical values calculated through circuital simulations, the values reported in Tables 1–3 have been derived for the coupling coefficients. By using (47) and (48) it is possible to calculate the expressions of the optimal loads. The values calculated for the three analyzed examples are summarized in Tables 1–3. With all the resonators the same quality factor, the resistive part of the loads is the same for all the receivers.

As per the imaginary parts, for the analyzed cases all the calculated values of $X_{LM,n}$ are negative, thus corresponding to load impedances with an inductor $L_{M,n}$ in series configurations with the resistive part $R_{LM,n}$.

Table 1. Parameters of the equivalent circuit and optimal loads of the WPT link illustrated in Figure 5a (Case 1).

L_n, ($n=0,1,2$) (nH)	C_n, ($n=0,1,2$) (pF)	Q	f_0 (MHz)
20.91	4.84	557	500

Coupling coefficients		
k_{01}	k_{02}	k_{12}
0.0215	0.0215	−0.0399

Optimal loads				
α	G_M	R_G (Ω)	$R_{LM,n}$, ($n=1,2$) (Ω)	$L_{LM,n}$, ($n=1,2$) (nH)
16.94	0.89	1.999	1.999	0.835

Table 2. Parameters of the equivalent circuit and optimal loads of the WPT link illustrated in Figure 5b (Case 2).

L_n, ($n=0,1,2,3$) (nH)	C_n, ($n=0,1,2,3$) (pF)	Q	f_0 (MHz)
20.91	4.84	557	500

Coupling coefficients					
k_{01}	k_{02}	k_{03}	k_{12}	k_{13}	k_{23}
0.0406	0.0406	0.0178	−0.0281	−0.0283	−0.0283

Optimal loads					
α	G_M	R_G (Ω)	$R_{LM,n}$, ($n=1,2,3$) (Ω)	$L_{LM,n}$, ($n=1,2$) (nH)	$L_{LM,3}$ (nH)
33.402	0.942	3.941	3.941	0.845	2.695

Table 3. Parameters of the equivalent circuit and optimal loads of the WPT link illustrated in Figure 5c (Case 3).

				L_n, ($n = 0,1,2,3,4$) (nH)	C_n, ($n = 0,1,2,3,4$) (pF)	Q	f_0 (MHz)			
				20.91	4.84	557	500			
					Coupling coefficients					
k_{01}	k_{02}	k_{03}	k_{04}	k_{12}	k_{13}	k_{14}	k_{23}		k_{24}	k_{34}
0.0123	0.0123	−0.0209	−0.0209	−0.0395	0.0084	0.0393	0.0393		0.0084	−0.0395
					Optimal loads					
		α	G_M	R_G (Ω)	$R_{LM,n}$, ($n = 1,2,3,4$) (Ω)	$L_{LM,n}$, ($n = 1,2$) (nH)	$L_{LM,j}$, ($j = 3,4$) (nH)			
		19.122	0.901	2.256	2.256	2.522	1.414			

The analytical values of the optimal loads have been validated through circuital simulations.

Two different sets of circuital simulations have been performed. A first set of simulations has been performed by modeling the links with lumped elements equivalent circuits with the parameters summarized in Tables 1–3; Figure 6 illustrates the circuit analyzed for case 1. The resistors R_n that appear in Figure 6 are related to the quality factors of the resonators, Q_n, through (29). A second set of simulations has been performed by modeling the analyzed links as $(N + 1)$–port black–box networks described by the impedance matrices provided by full–wave simulations, being N the number of receivers (i.e., $N = 2$ for case 1, $N = 3$ for case 2, $N = 4$ for case 3). In more detail, referring to case 1, simulations have been performed by replacing the network in the dashed square of Figure 6 with a three–port black-box component described by the impedance matrix of the link calculated through circuital simulations.

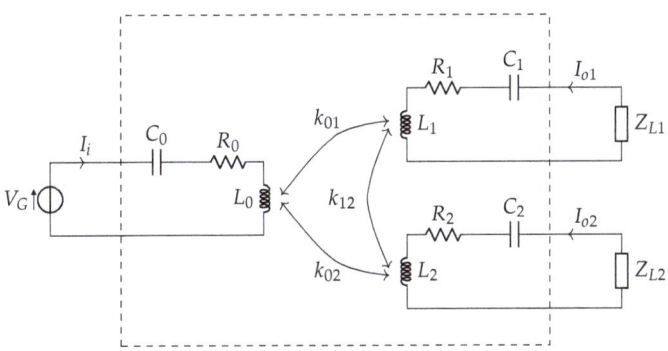

Figure 6. Equivalent circuit analyzed for Case 1.

First, the optimal values provided by the theory for the resistive parts of the loads have been validated. Simulations have been performed by varying the resistive part of the loads, the values calculated for G_P are given in Figure 7a–c.

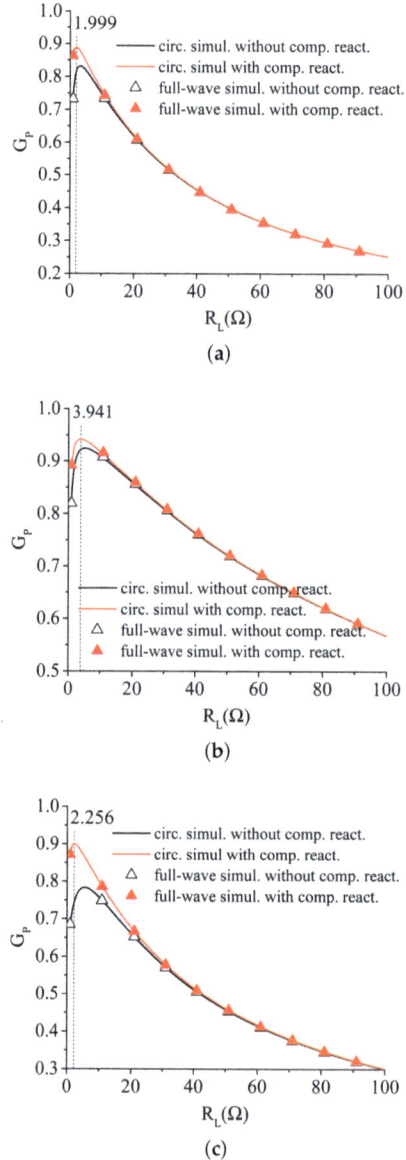

Figure 7. Power gain calculated through circuital simulations by varying the resistive part of the loads R_{Ln}. (**a**) Case 1, the link has two receivers with $R_{L1} = R_{L2} = R_L$; (**b**) Case 2, the link has three receivers with $R_{L1} = R_{L2} = R_{L3} = R_L$; (**c**) case 3, the link has four receivers with $R_{L1} = R_{L2} = R_{L3} = R_{L4} = R_L$. The figure compares full-wave and circuital simulation results obtained for the case of purely resistive loads and for the case of loads with the compensating inductances given in Tables 1–3.

The results obtained for the case of purely resistive loads (i.e., $Z_{Ln} = R_L$) are compared with those obtained for the case of loads with the compensating inductances given in Table 3, i.e., for $Z_{Ln} = R_L + j\omega L_{LM,n}$. In the figures, the triangles have been used for the results obtained by modeling the link with the impedance matrix provided by full-wave simulations; however, the solid lines have been used for the results obtained by modeling the links with the lumped elements equivalent circuit. It can be seen that the results obtained for the two representations of the links are coincident.

As per the optimal values of R_L, in each figure the value of R_L for which G_P is maximized is highlighted by a dashed vertical line. It can be seen that the values calculated through circuital simulations confirm the theoretical values. Finally, with regard to the compensating inductances $L_{LM,n}$, it seems that they play a more or less important role in maximizing G_P depending on the analyzed case. For instance, according to the achieved results the compensating reactances play a marginal role in maximizing G_P for case 2 while they seem to be more relevant for case 3. However, for all the three analyzed cases it is confirmed that their presence allows obtaining the maximum value of G_P provided by the theory.

The behavior of the power delivered to the loads as function of the generator impedance has been also investigated. Simulations have been performed by using as source a voltage generator with a series impedance R_G. The total output power has been calculated by varying R_G when the loads assume the optimal values provided by the theory. The results are given in Figure 8a–c, data obtained for $Z_{Ln} = R_{LM,n}$ and $Z_{Ln} = R_{LM,n} + j\omega L_{LM,n}$ are compared. The dashed vertical lines highlight the values of R_G maximizing P_o for the case $Z_{Ln} = R_{LM,n} + j\omega L_{LM,n}$. It can be verified that these values are in a perfect agreement with the optimal values provided by the theory. As it can be seen, the use of the generator impedance provided by (43) allows maximizing the power delivered to the loads when they are set to maximize G_P. Additionally, it can be seen that the compensation reactances are crucial to maximize the power transferred to the loads.

It is worth observing that the output power illustrated in Figure 8a–c is the total output power delivered (i.e., the sum of the power delivered to the loads) when the network operates at maximum G_P. From the figures it is evident that if R_G is not optimized, although the network operates with efficiency values close to one, only a small portion of the power available from the generator is delivered to the load.

Finally, the case of a link consisting of three coils with different dimensions has been analyzed (case 4). The geometry analyzed through full-wave simulations is illustrated in Figure 9. All the coils have been designed by using a copper wire with a radius of 0.3 mm. The radius of the transmitting coil is 10 mm, those of the first and second receivers are 7.5 mm and 5 mm, respectively. Also, in this case an operating frequency f_0 of 500 MHz has been assumed.

The impedance matrix as calculated from full-wave simulations is:

$$Z_{Case4} = \begin{pmatrix} 0.229 & -10.08\,j & 8.66\,j \\ -10.08\,j & 0.145 & 2.82\,j \\ 8.66\,j & 2.82\,j & 0.118 \end{pmatrix}. \qquad (52)$$

The parameters derived for the equivalent circuit are summarized in Table 4.

The simulated results obtained for G_P are illustrated in Figure 10. Circuital simulations have been performed by modeling the link as a three–port black–box component described by the impedance matrix calculated through full–wave simulations. In this case, the two receivers have slightly different values of the quality factors; accordingly, the theory predicts slightly different values for $R_{LM,1}$ and $R_{LM,2}$. To verify the expected optimal values, simulations have been performed by terminating the receivers ports on the impedances $Z_{L1} = R_{L1} + j\omega L_{LM,1}$ and $Z_{L2} = R_{L2} + j\omega L_{LM,2}$ and by varying both $R_{L,1}$ and $R_{L,2}$. From Figure 10 it can be seen that circuital simulations confirm the theory, a maximum of about 0.98 is obtained for G_P when $R_{L1} = 11.07\,\Omega$ and $R_{L2} = 9.01\,\Omega$; however, from the figure it can also be seen that values of G_P very close to its maximum (i.e., values greater than 0.97) are obtained for a wide range of values of R_{L1} and R_{L2}.

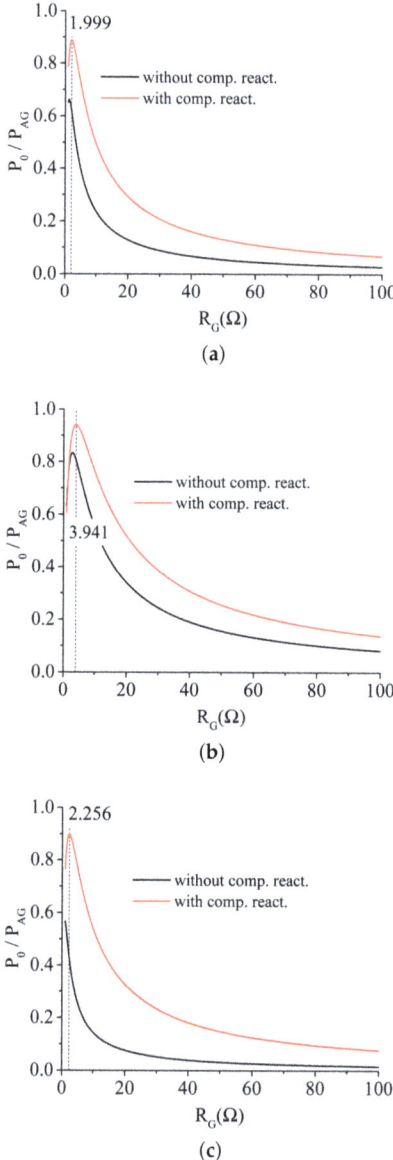

Figure 8. Total output power as function of the generator impedance R_G corresponding to the optimal loads $Z_{LM,n} = R_{LM,n} + j\,X_{LM,n}$; results obtained with and without the compensating reactances $X_{LM,n}$. (**a**) Case 1; (**b**) Case 2; (**c**) Case 3.

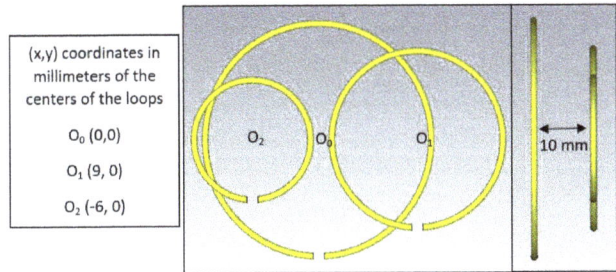

Figure 9. Case 4: WPT link analyzed through full-wave simulations. The link has a single transmitter and two receivers.

Table 4. Parameters of the equivalent circuit and optimal loads of the WPT link illustrated in Figure 9 (Case 4).

L_0 (nH)	L_1 (nH)	L_2 (nH)	C_0 (pF)	C_1 (pF)	C_2 (pF)	Q_0	Q_1	Q_2	f_0 (MHz)
45.38	28.67	20.91	2.23	3.53	4.84	622	621	557	500
			\multicolumn{3}{c}{Coupling coefficients}				
			k_{01}	k_{02}	k_{12}				
			−0.089	0.0894	0.0367				
			\multicolumn{3}{c}{Optimal loads}				
	α	G_M	R_G (Ω)	$R_{LM,1}$ (Ω)	$R_{LM,2}$ (Ω)	$L_{LM,1}$ (nH)	$L_{LM,2}$ (nH)		
	76.38	0.974	17.49	11.07	9.01	0.948	0.852		

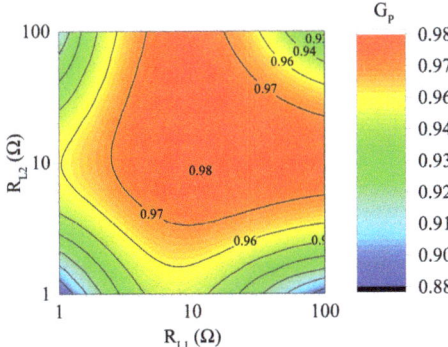

Figure 10. Power gain calculated through circuital simulations for case 4. Results obtained by varying the resistive part of the loads when the reactive parts are set according to the optimal values provided by the theory (see Table 4).

Finally, the behavior obtained for the total output power P_o as function of the generator impedance is illustrated in Figure 11. In this case, simulations have been performed by terminating the ports of the receivers on the optimal load impedances, i.e., $Z_{L1} = R_{LM,1} + j\omega L_{LM,1}$ and $Z_{L2} = R_{LM,2} + j\omega L_{LM,2}$. As for the previously analyzed cases, simulations confirm the importance of suitably selecting the generator impedance for maximizing P_o when the loads are those maximizing G_P.

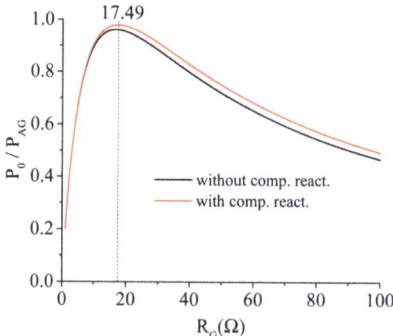

Figure 11. Total output power calculated for case 4 by varying the generator impedance when the loads are those maximizing G_P.

5. Conclusions

In this paper, the case of a WPT link using a Single-Input Multiple-Output (SIMO) configuration is analyzed. The solution for maximizing the efficiency is derived from a generalized eigenvalue problem. The main advantage of the presented theory is its complete generality, no specific assumptions are made about the link. In fact, the proposed approach is valid for any strictly passive and reciprocal $(N+1)$–port network. Additionally, the desired solution can be derived directly from the impedance matrix with simple algebraic operations. Theoretical formulas for the optimal loads maximizing the efficiency are derived for the case of a resonant inductive link with a generic number N of possibly coupled receivers.

The results obtained this way for the optimal loads are coincident with those reported in the previous literature were the first-order condition on the partial derivatives of the efficiency has been exploited in order to find the desired solution. As a further validation of the derived formulas, several numerical examples have been analyzed through full–wave and circuital simulations. In particular, four different links have been considered. The first three analyzed links use identical resonators for the transmitter and the receivers and differ for the number of receivers. The last analyzed case is a link using two receivers; in this case, the three resonators have been designed so to have different values of the equivalent inductance. More specifically, the transmitter has been designed so to have a larger inductance with respect to the receivers, so to obtain higher values of the couplings with respect to the previously analyzed cases. For all the investigated cases, the impedance matrix has been calculated through full-wave simulations and the presented theory applied so to determine the maximum realizable efficiency and the optimal terminating impedances. The correctness of the analytical data has been verified through simulations performed for evaluating the efficiency of the links. According to the theoretical data, simulations confirm that the maximum realizable efficiency of a resonant inductive link in SIMO configuration does not depend on the coupling among the receiving resonators. In fact, it is demonstrated that a possible coupling among the receivers can be always compensated by using suitable complex loads.

The possibility of maximizing the power delivered to the loads when they are set to maximize the efficiency has been also discussed. In fact, the presented theory provides both:

- the expressions of the loads that maximize the efficiency,
- the expression of the generator impedance that allows maximizing the power entering the network when the loads are those maximizing the efficiency.

Maximizing the power entering the network for a given efficiency corresponds maximizing the power delivered to the loads. The reported results highlight the importance of also optimizing the generator impedance to avoid that. Despite the high efficiency values, only a small portion of the power available from the generator is transferred to the loads.

As future developments of the presented research, experimental tests will be performed to verify the application of the proposed theory to a real application. Furthermore, in a future work the analysis presented in this paper will be extended to a generic MIMO (Multiple Input Multiple-Output) system.

Author Contributions: Conceptualization, methodology, validation, G.M., M.M.; writing—original draft preparation, G.M., M.M.; writing—review and editing, B.M., A.C. and L.T. All authors have read and agreed to the published version of the manuscript.

Funding: This research received no external funding.

Acknowledgments: The authors would like to remember the colleague Franco Mastri who suddenly passed away on 3 April 2020. He was a great colleague and a profound scientist. Fundamental discussions and studies on the theoretical modeling of near-field WPT systems were of great inspiration also for the results presented in this work.

Conflicts of Interest: The authors declare no conflict of interest.

Appendix A. How to Solve the Generalized Eigenvalue Problem

The following generalized eigenvalue problem is considered:

$$\mathbf{A}\mathbf{x} = \lambda \mathbf{B}\mathbf{x}. \tag{A1}$$

In this Appendix the procedure for calculating the eigenvalues λ and the corresponding eigenvectors $\mathbf{x} = \mathbf{I}$ will be illustrated. By introducing the variable η:

$$\eta = \frac{1}{\lambda} \tag{A2}$$

it is possible to write:

$$(\mathbf{B} - \eta \mathbf{A})\,\mathbf{x} = 0. \tag{A3}$$

By using (18) and (19), it is possible to obtain:

$$\mathbf{B} - \eta \mathbf{A} = \left[\begin{array}{c|c} \tilde{\mathbf{Z}}_{ii} & \mathbf{Z}_{io} + \eta \mathbf{Z}_{oi}^{\dagger} \\ \hline \mathbf{Z}_{io}^{\dagger} + \eta \mathbf{Z}_{oi} & \eta \tilde{\mathbf{Z}}_{oo} \end{array} \right], \tag{A4}$$

where the following definitions have been introduced:

$$\begin{aligned}\tilde{\mathbf{Z}}_{ii} &= \mathbf{Z}_{ii} + \mathbf{Z}_{ii}^{*} \\ \tilde{\mathbf{Z}}_{oo} &= \mathbf{Z}_{oo} + \mathbf{Z}_{oo}^{\dagger}.\end{aligned} \tag{A5}$$

The eigenvalues can be obtained from:

$$\det(\mathbf{B} - \eta \mathbf{A}) = 0. \tag{A6}$$

Considering that for a matrix $\mathbf{M}$ partitioned in 4 submatrix $\mathbf{M}_{11}, \mathbf{M}_{12}, \mathbf{M}_{21}, \mathbf{M}_{22}$:

$$\mathbf{M} = \left[\begin{array}{c|c} \mathbf{M}_{11} & \mathbf{M}_{12} \\ \hline \mathbf{M}_{21} & \mathbf{M}_{22} \end{array} \right], \tag{A7}$$

the determinant is given by:

$$\det(\mathbf{M}) = \det(\mathbf{M}_{11} - \mathbf{M}_{12}\mathbf{M}_{22}^{-1}\mathbf{M}_{21})\det(\mathbf{M}_{22}), \tag{A8}$$

it is possible to write
$$\det(\mathbf{B} - \eta \mathbf{A}) = -\eta^{N-1}(c_0 \eta^2 + c_1 \eta + c_2) \det(\tilde{\mathbf{Z}}_{oo}). \tag{A9}$$

where the coefficients c_0, c_1 and c_2 are given by:

$$\begin{aligned}
c_0 &= \mathbf{Z}_{oi}^{\dagger} \tilde{\mathbf{Z}}_{oo}^{-1} \mathbf{Z}_{oi}, \\
c_1 &= -\tilde{\mathbf{Z}}_{ii} + \mathbf{Z}_{io} \tilde{\mathbf{Z}}_{oo}^{-1} \mathbf{Z}_{oi} + \mathbf{Z}_{oi}^{\dagger} \tilde{\mathbf{Z}}_{oo}^{-1} \mathbf{Z}_{io}^{\dagger}, \\
c_2 &= \mathbf{Z}_{io} \tilde{\mathbf{Z}}_{oo}^{-1} \mathbf{Z}_{oi}^{\dagger}.
\end{aligned} \tag{A10}$$

Accordingly, the non-trivial solutions are provided by the equation:

$$c_2 \lambda^2 + c_1 \lambda + c_0 = 0. \tag{A11}$$

References

1. Karalis, A.; Joannopoulos, J.D.; Soljačić, M. Efficient wireless non-radiative mid-range energy transfer. *Ann. Phys.* **2008**, *323*, 34–48. [CrossRef]
2. Monti, G.; Paolis, M.V.D.; Corchia, L.; Tarricone, L. Wireless Resonant Energy link for Pulse Generators Implanted in the Chest. *IET Microw. Antennas Propag.* **2017**, *11*, 2201–2210. [CrossRef]
3. Carvalho, N.B.; Georgiadis, A.; Costanzo, A.; Stevens, N.; Kracek, J.; Pessoa, L.; Rogier, H. Europe and the future for WPT. *IEEE Microw. Mag.* **2017**, *18*, 56–87. [CrossRef]
4. Rim, C.T.; Mi, C. *Wireless Power Transfer for Electric Vehicles and Mobile Devices*; Wiley-IEEE Press: Hoboken, NJ, USA, 2017.
5. Inagaki, N. Theory of Image Impedance Matching for Inductively Coupled Power Transfer Systems. *IEEE Trans. Microw. Theory Tech.* **2014**, *62*, 901–908. [CrossRef]
6. Mastri, F.; Mongiardo, M.; Monti, G.; Dionigi, M.; Tarricone, L. Gain expressions for resonant inductive wireless power transfer links with one relay element. *Wirel. Power Transf.* **2018**, *5*, 27–41. [CrossRef]
7. Monti, G.; Costanzo, A.; Mastri, F.; Mongiardo, M. Optimal design of a wireless power transfer link using parallel and series resonators. *Wirel. Power Transf.* **2016**, *3*, 105–116. [CrossRef]
8. Yoon, I.; Ling, H. Investigation of near-field wireless power transfer under multiple transmitters. *IEEE Antenna Wirel. Propag. Lett.* **2011**, *10*, 662–665. [CrossRef]
9. Pacini, A.; Costanzo, A.; Aldhaher, S.; Mitcheson, P.D. Load- and Position-Independent Moving MHz WPT System Based on GaN-Distributed Current Sources. *IEEE Trans. Microw. Theory Tech.* **2017**, *65*, 5367–5376. [CrossRef]
10. Lang, H.D.; Ludwig, A.; Sarris, C.D. Convex Optimization of Wireless Power Transfer Systems with Multiple Transmitters. *IEEE Trans. Antennas Propag.* **2014**, *62*, 4623–4636. [CrossRef]
11. Monti, G.; Che, W.; Wang, Q.; Costanzo, A.; Dionigi, M.; Mastri, F.; Mongiardo, M.; Perfetti, R.; Tarricone, L.; Chang, Y. Wireless Power Transfer with Three-Ports Networks: Optimal Analytical Solutions. *IEEE Trans. Circuits Syst.* **2017**, *62*, 494–503. [CrossRef]
12. Arakawa, T.; Goguri, S.; Krogmeier, J.V.; Kruger, A.; Love, D.J.; Mudumbai, R.; Swabey, M.A. Optimizing wireless power transfer from multiple transmit coils. *IEEE Access* **2018**, *6*, 23828–23838. [CrossRef]
13. Kim, G.; Boo, S.; Kim, S.; Lee, B. Control of Power Distribution for Multiple Receivers in SIMO Wireless Power Transfer System. *J. Electromagn. Eng. Sci.* **2018**, *18*, 221–230. [CrossRef]
14. Cai, W.; Ma, D.; Tang, H.; Lai, X.; Liu, X.; Sun, L. Highly Efficient Target Power Control for Two-Receiver Wireless Power Transfer Systems. *Energies* **2018**, *11*, 2726. [CrossRef]
15. Liu, M.; Fu, M.; Wang, Y.; Ma, C. Battery Cell Equalization via Megahertz Multiple-Receiver Wireless Power Transfer. *IEEE Trans. Power Electron.* **2018**, *33*, 4135–4144. [CrossRef]
16. Cai, W.; Lai, X.; Ma, D.; Tang, H.; Hashmi, K.; Xu, J. Management of Multiple-Transmitter Multiple-Receiver Wireless Power Transfer Systems Using Improved Current Distribution Control Strategy. *Electronics* **2019**, *8*, 1160. [CrossRef]
17. Luo, C.; Qiu, D.; Lin, M.; Zhang, B. Circuit Model and Analysis of Multi-Load Wireless Power Transfer System Based on Parity-Time Symmetry. *Energies* **2020**, *13*, 3260. [CrossRef]

18. Wagih, M.; Komolafe, A.; Zaghari, B. Dual-Receiver Wearable 6.78 MHz Resonant Inductive Wireless Power Transfer Glove Using Embroidered Textile Coils. *IEEE Access* **2020**, *8*, 24630–24642. [CrossRef]
19. Ahn, D.; Kim, S.; Kim, S.; Moon, J.; Cho, I. Wireless Power Transfer Receiver with Adjustable Coil Output Voltage for Multiple Receivers Application. *IEEE Trans. Ind. Electron.* **2019**, *66*, 4003–4012. [CrossRef]
20. Vu, V.; Phan, V.; Dahidah, M.; Pickert, V. Multiple Output Inductive Charger for Electric Vehicles. *IEEE Trans. Power Electron.* **2019**, *34*, 7350–7368. [CrossRef]
21. Monti, G.; Dionigi, M.; Mongiardo, M.; Perfetti, R. Optimal Design of Wireless Energy Transfer to Multiple Receivers: Power Maximization. *IEEE Trans. Microw. Theory Tech.* **2017**, *65*, 260–269. [CrossRef]
22. Fu, M.; Yin, H.; Liu, M.; Wang, Y.; Ma, C. A 6.78 MHz Multiple-Receiver Wireless Power Transfer System with Constant Output Voltage and Optimum Efficiency. *IEEE Trans. Power Electron.* **2018**, *33*, 5330–5340. [CrossRef]
23. Lin, H.; Li, L. Efficiency Analysis and Optimization for Multiple-Receiver Magnetic Coupling Resonant Wireless Power Transfer System. In Proceedings of the 2020 5th International Conference on Computer and Communication Systems (ICCCS), Shanghai, China, 22–24 February 2020; pp. 742–747.
24. Lee, K.; Chae, S.H. Comparative Analysis of Frequency-Selective Wireless Power Transfer for Multiple-Rx Systems. *IEEE Trans. Power Electron.* **2020**, *35*, 5122–5131. [CrossRef]
25. Fu, M.; Zhang, T.; Ma, C.; Zhu, X. Efficiency and Optimal Loads Analysis for Multiple-Receiver Wireless Power Transfer Systems. *IEEE Trans. Microw. Theory Tech.* **2015**, *63*, 3463–3477. [CrossRef]
26. Fu, M.; Zhang, T.; Zhu, X.; Luk, P.C.; Ma, C. Compensation of Cross Coupling in Multiple-Receiver Wireless Power Transfer Systems. *IEEE Trans. Ind. Inform.* **2016**, *12*, 474–482. [CrossRef]
27. Ishihara, M.; Fujiki, K.; Umetani, K.; Hiraki, E. Automatic Active Compensation Method of Cross-Coupling in Multiple-receiver Resonant Inductive Coupling Wireless Power Transfer Systems. In Proceedings of the 2019 IEEE Energy Conversion Congress and Exposition (ECCE), Baltimore, MD, USA, 29 September–3 October 2019; pp. 4584–4591.
28. Duong, Q.T.; Okada, M. Maximum efficiency formulation for inductive power transfer with multiple receivers. *IEICE Electron. Exp.* **2016**, *22*, 20160915. [CrossRef]
29. Yuan, Q.; Aoki, T. Practical applications of universal approach for calculating maximum transfer efficiency of MIMO-WPT system. *Wirel. Power Transf.* **2020**, *7*, 86–94. [CrossRef]
30. Duong, Q.; Okada, M. Maximum Efficiency Formulation for Multiple-Input Multiple-Output Inductive Power Transfer Systems. *IEEE Trans. Microw. Theory Tech.* **2018**, *66*, 3463–3477. [CrossRef]
31. Desoer, C. The maximum power transfer theorem for n-ports. *IEEE Trans. Circuit Theory* **1973**, *20*, 328–330. [CrossRef]

© 2020 by the authors. Licensee MDPI, Basel, Switzerland. This article is an open access article distributed under the terms and conditions of the Creative Commons Attribution (CC BY) license (http://creativecommons.org/licenses/by/4.0/).

Article

Capacitive Wireless Power Transfer with Multiple Transmitters: Efficiency Optimization

Ben Minnaert [1,*], Alessandra Costanzo [2], Giuseppina Monti [3] and and Mauro Mongiardo [4]

1. Department of Industrial Science and Technology, Odisee University College of Applied Sciences, 9000 Ghent, Belgium
2. Department of Electrical, Electronic and Information Engineering Guglielmo Marconi, University of Bologna, 40126 Bologna, Italy; alessandra.costanzo@unibo.it
3. Department of Engineering for Innovation, University of Salento, 73100 Lecce, Italy; giuseppina.monti@unisalento.it
4. Department of Engineering, University of Perugia, 06123 Perugia, Italy; mauro.mongiardo@unipg.it
* Correspondence: ben.minnaert@odisee.be

Received: 28 May 2020; Accepted: 28 June 2020; Published: 6 July 2020

Abstract: Wireless power transfer with multiple transmitters can have several advantages, including more robustness against misalignment and extending the mobility and range of the receiver(s). In this work, the efficiency maximization problem is analytically solved for a capacitive wireless power transfer system with multiple coupled transmitters and a single receiver. It is found that the system efficiency can be increased by adding more transmitters. Moreover, it is proven that the cross-coupling between the transmitters can be eliminated by adding shunt susceptances at the input ports. Optimal values for the input currents and receiver load are determined to achieve maximum efficiency. As well the optimal load, the optimal input currents and the maximum efficiency are independent on the cross-coupling. By impedance-matching the internal conductances of the generators, the maximum-efficiency solution also becomes the one that provides the maximum output power. Finally, by expressing each transmitter–receiver link with its kQ-product, the maximum system efficiency can be calculated. The analytical results are verified by circuital simulation.

Keywords: capacitive wireless power transfer; resonance; wireless power transfer; power-transfer efficiency; multiports; multiple-input single-output

1. Introduction

Near-field wireless power transfer (WPT) represents a promising solution for wirelessly providing power to electronic devices. Two main technologies exist: inductive and capacitive WPT, based on resonant magnetic or electric coupling, respectively. The simplest setup consists of two resonators: a transmitting resonator, powered by an input supply, which transfers power wirelessly to a receiving resonator connected to the load (SISO: single-input single-output).

For certain applications, WPT from multiple transmitters to a single receiver (MISO: multiple-input single-output) can be beneficial compared to the single-transmitter configuration:

- First, multiple transmitters targeting a single receiver results in a certain robustness against misalignment or mispositioning of the receiver. This extends the mobility of the receiver. If, for example, multiple transmitters are present in a planar configuration, the power transfer can be realized for a wide possibility of planar receiver positions.
- Second, the above implies that the use of multiple transmitters can extend the range of the wireless power transfer.
- A decentralization of the transmitters can possibly facilitate high-power energy transfer by applying multiple cheaper low-power input supplies.

- Multiple transmitters make the WPT system less vulnerable to foreign blocking objects.
- Last but not least, as will be shown in this work, multiple transmitters allow for a higher system efficiency for given coupling coefficients since they allow for a higher degree of freedom for the power input distribution.

In this work, focus will lie on capacitive power transfer (CPT), which utilizes a high-frequency electric field as a medium to transfer energy wirelessly. Main advantages compared to inductive WPT are low-weight and cost, a minimal eddy-current loss, and a larger robustness against misalignment. A typical CPT coupler consists of four metal plates: two plates at the transmitter side, and two at the receiver side, resulting in a return path for the current [1]. CPT has been demonstrated in low-power applications such as portable electronics [2,3], integrated circuits [4], drones [5], and biomedical devices [6,7]. However, also at higher power levels, up to several kW [8], CPT can be applied—e.g., in automatic guided vehicles [9] and electric vehicles [10–12].

Maximizing the efficiency of an *inductive* WPT system with multiple transmitters has already been solved, e.g., [13–18]. Additionally, for the general WPT system, this problem has been solved by reducing the entire system to an impedance matrix of a multiport network [19–21]. However, this methodology loses the internal structure of the WPT system, e.g., the coupling strengths between transmitters and receiver. Moreover, a CPT system can be more easily described by its admittance matrix instead of its impedance matrix.

Efficiency maximization for CPT was already solved for a single transmitter with multiple receivers (SIMO: single-input multiple-output) [22,23], but to date, an analysis specifically for a CPT system with multiple (coupled) transmitters (MISO) is lacking.

In this work, a CPT transfer system with *any* number of transmitters and a single receiver is considered. Varying the receiver's loads (e.g., via impedance matching) and/or the input currents results in different values for the power-transfer efficiency (also called power gain) of the CPT system. In this work, the load and input currents that maximize the power-transfer efficiency are determined while taking into account, among others, the coupling strengths between transmitters and receiver. More specifically, the contributions are the following:

- After constructing the general CPT system with N transmitters and a single receiver (Section 2), the input power, output power, and efficiency as functions of the characteristics of the network are determined (Section 3).
- The optimal current–voltage relationships at the transmitter and receiver ports are calculated (Section 4). From these relationships, closed-form expressions for the optimal load, input current ratios, and the maximum efficiency are analytically determined (Section 4).
- By matching the internal shunt admittance of the generators to the system, the maximum-efficiency solution coincides with the configuration that maximizes the output power (Section 4).
- It is shown that the cross-coupling between the transmitters does not influence the value of the maximum efficiency or optimal load: by including a reactive part at the transmitter side, the impact of cross-coupling can be neutralized. Moreover, the efficiency of the CPT system can be increased by adding extra transmitters (Section 5).
- It is demonstrated that the maximum efficiency of the multiple transmitter system can be estimated by measuring the individual transmitter–receiver links (Section 5).
- Finally, the analytical solution is demonstrated on an example equivalent circuit of a CPT system. The results are verified by numerical circuit simulation for a system with three transmitters and a single receiver (Section 6).

2. Problem Description

Figure 1 depicts a CPT system with N transmitters (on the bottom, subscripts 1 to N) and a single receiver (on top, subscript 0). The transmitters are powered by a power supply control circuit. Each transmitter can operate at a different voltage and phase, but the operating angular frequency ω_0 is the same for all transmitters. The resistive and reactive components within each transmitter are described by the conductances g_{nn} and susceptances b_{nn} ($n = 1, \ldots, N$), respectively. In the remainder of this work, the subscript n always counts from 1 to N.

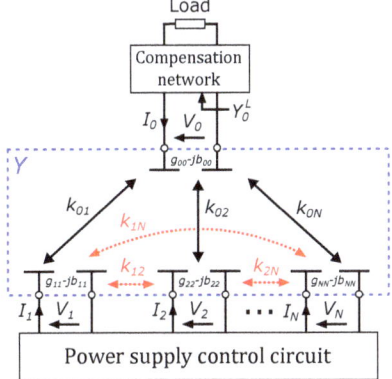

Figure 1. A general capacitive wireless power transfer system with N transmitters (bottom) and a single receiver (top). The (desired) electric couplings between transmitters and receiver are depicted by the full arrows. The undesired cross-couplings between the transmitters themselves are indicated by the dashed arrows.

Energy is transferred wirelessly to the load of the receiver, represented by the admittance Y_0^L (including a possible compensation circuit). The conductance g_{00} and susceptance b_{00} correspond to the resistive and reactive part of the receiving resonator, respectively.

The strength of the electric coupling between each transmitter and the receiver is given by the coupling factor k_{0n}, a dimensionless number which can vary from zero (no coupling) to unity (maximum coupling). In a practical CPT system, the electric coupling between the transmitters and receiver is desired to realize wireless power transmission. However, an undesired (nonzero) electric cross-coupling can be present between the transmitters themselves, represented by k_{nm} ($n, m = 1, \ldots, N; n \neq m$). The coupling factor is defined as [24,25]

$$k_{ij} = \frac{C_{ij}}{\sqrt{C_i C_j}}, \tag{1}$$

for $i, j = 0, \ldots, N; i \neq j$, where C_n is the transmitter capacitance of the n-th transmitter, C_0 is the receiver capacitance, and C_{ij} is the mutual capacitance, corresponding to the electric coupling. Note that C_0 and C_n do not correspond to the capacitance between the physical transmitter and receiver plate, but to an equivalent circuit representation of electric coupling [25]. The measurement procedure to determine the value of these capacitances is described in [24].

The CPT system can be considered as a multiport with N input ports (the N transmitters) and one output port (the receiver). The multiport is indicated by the dashed rectangle in Figure 1. Notice that this $(N+1)$-port network is linear and reciprocal due to the passive components it is constructed from. The currents through and voltages at the $(N+1)$ ports are given by the peak current phasors I_j and peak voltage phasors V_j, as defined in Figure 1 ($j = 0, \ldots, N$).

The problem description is the following: given the network of Figure 1 (with given and fixed values for the components of the CPT network and coupling factors), determine the values of the load admittance Y_0^L and input currents I_n that maximize the power-transfer efficiency η. The problem can be reduced to finding the current and voltage phasors at the ports, corresponding to the optimal efficiency configuration. Therefore, any remote electronics external to the wireless link (e.g., rectifiers, matching networks, actual passive loads,...) can be ignored, since it can be taken into account once the optimal current–voltage relationship at the ports are determined.

3. Power and Efficiency of the CPT System

First, the (normalized) admittance matrix, input power, output power, and efficiency will be expressed as functions of the characteristics of the network. The efficiency is not yet maximized in this section.

3.1. Admittance Matrix

The CPT system can be fully characterized by its admittance matrix Y. The admittance matrix Y describes the relation between the port currents and port voltages:

$$I = Y \cdot V, \tag{2}$$

with the $(N+1) \times 1$ matrices V and I defined as

$$V = \begin{bmatrix} V_1 \\ V_2 \\ \vdots \\ V_N \\ V_0 \end{bmatrix}, I = \begin{bmatrix} I_1 \\ I_2 \\ \vdots \\ I_N \\ I_0 \end{bmatrix}. \tag{3}$$

The admittance matrix Y of the CPT system can be written as [25]

$$Y = \begin{bmatrix} g_{11} - jb_{11} & -jb_{12} & \cdots & -jb_{1N} & -jb_{10} \\ -jb_{21} & g_{22} - jb_{22} & \cdots & -jb_{2N} & -jb_{20} \\ \vdots & \vdots & \ddots & \vdots & \vdots \\ -jb_{N1} & -jb_{N2} & \cdots & g_{NN} - jb_{NN} & -jb_{N0} \\ -jb_{01} & -jb_{02} & \cdots & -jb_{0N} & g_{00} - jb_{00} \end{bmatrix}, \tag{4}$$

with $b_{ij} = \omega_0 C_{ij}$, $(i,j = 0,\ldots,N; i \neq j)$. Since the network is reciprocal, Y is symmetric: $b_{ij} = b_{ji}$. In practical applications, the admittance matrix Y can be measured. Note that each transmitter and the receiver have a self-susceptance expressed by $-b_{jj}$.

A normalization matrix n is defined:

$$n = \begin{bmatrix} \frac{1}{\sqrt{\omega_0 C_1}} & \cdots & 0 & 0 \\ \vdots & \ddots & \vdots & \vdots \\ 0 & \cdots & \frac{1}{\sqrt{\omega_0 C_N}} & 0 \\ 0 & \cdots & 0 & \frac{1}{\sqrt{\omega_0 C_0}} \end{bmatrix}, \tag{5}$$

in order to normalize the admittance matrix:

$$y = n \cdot Y \cdot n = \begin{bmatrix} \frac{1}{Q_1} - jk_{11} & \cdots & -jk_{1N} & -jk_{10} \\ \vdots & \ddots & \vdots & \vdots \\ -jk_{N1} & \cdots & \frac{1}{Q_N} - jk_{NN} & -jk_{N0} \\ -jk_{01} & \cdots & -jk_{0N} & \frac{1}{Q_0} - jk_{00} \end{bmatrix}, \tag{6}$$

with quality factor Q_i of the coupled resonators $(i,j = 0,\ldots,N)$:

$$Q_i = \frac{\omega_0 C_i}{g_{ii}} \tag{7}$$

and

$$k_{ij} = \frac{b_{ij}}{\omega_0 \sqrt{C_i C_j}}. \tag{8}$$

For $i \neq j$, the parameter k_{ij} corresponds to the coupling factor between circuits i and j.
The voltages and currents are normalized as follows:

$$i = n \cdot I, \tag{9}$$

$$v = n^{-1} \cdot V. \tag{10}$$

The normalized current–voltage relationship is thus given by

$$i = y \cdot v. \tag{11}$$

The real and imaginary parts of the (normalized) current and voltage phasors can be explicitly written out: $i_n = i_n^{re} + j i_n^{im}$ and $v_n = v_n^{re} + j v_n^{im}$. Without loss of generality, we choose v_0 as the reference phasor, i.e., $v_0^{re} = v_0, v_0^{im} = 0$.

3.2. Input Power

The input power P_n ($n = 1, \ldots, N$) for the n-th transmitter system is given by

$$P_n = \frac{1}{2} \Re(v_n i_n^*), \tag{12}$$

where i_n^* is the complex conjugate of i_n, and $\Re(v_n i_n^*)$ is the real part of $v_n i_n^*$. This result for P_n is

$$P_n = \frac{1}{2}(v_n^{re} i_n^{re} + v_n^{im} i_n^{im}). \tag{13}$$

The total input power P_{in} of the entire CPT system is

$$P_{in} = \sum_{n=1}^{N} P_n. \tag{14}$$

Substituting the currents from Equation (11) into the above equation results in the total input power P_{in}:

$$P_{in} = \frac{1}{2} \sum_{n=1}^{N} \frac{1}{Q_n} [(v_n^{re})^2 + (v_n^{im})^2] + \frac{1}{2} \sum_{n=1}^{N} k_{n0} v_0 v_n^{im}. \tag{15}$$

The input power P_{in} is expressed as function of the parameters of the network and the port voltages.

3.3. Output Power

Analogously, the output power can be determined as a function of the network variables and port voltages.

Applying the passive sign convention, the output power P_{out} can be written as

$$P_{out} = -\frac{1}{2} \Re(v_0 i_0^*) = -\frac{1}{2} v_0 i_0^{re}. \tag{16}$$

Substituting the currents from Equation (11) into the above equation, the normalized output power P_{out} is determined:

$$P_{out} = -\frac{1}{2Q_0} v_0^2 - \frac{1}{2} v_0 \sum_{n=1}^{N} k_{0n} v_n^{im}. \tag{17}$$

3.4. Efficiency

The power-transfer efficiency η or power gain of the CPT system is defined as

$$\eta = \frac{P_{out}}{P_{in}}, \tag{18}$$

with the expressions for P_{in} and P_{out} given by Equations (15) and (17). Hence, η is written as a function of the parameters of the circuit and the port voltages.

4. Maximum-Efficiency Solution

In the previous section, the general expressions for input power, output power and efficiency were found. Now, the configuration at maximum power-transfer efficiency will be considered. The notation η_{max} is applied in this section to indicate that the power-transfer efficiency equals its maximum attainable value.

4.1. First-Order Necessary Condition

To find the output voltages at the maximum system efficiency configuration, the first-order necessary condition is applied to generate a system of $2N$ equations [20,26]:

$$\frac{\partial \eta}{\partial v_n^{re}} = 0, \tag{19}$$

$$\frac{\partial \eta}{\partial v_n^{im}} = 0. \tag{20}$$

The optimal input voltages v_n are the solution of the above system. Unfortunately, solving the system directly is not straightforward. First, the quotient rule for derivatives is applied. With Equation (18), the system becomes

$$P_{in} \frac{\partial P_{out}}{\partial v_n^{re}} - P_{out} \frac{\partial P_{in}}{\partial v_n^{re}} = 0, \tag{21}$$

$$P_{in} \frac{\partial P_{out}}{\partial v_n^{im}} - P_{out} \frac{\partial P_{in}}{\partial v_n^{im}} = 0. \tag{22}$$

Substituting the derivatives of Equations (15) and (17) to v_n^{re} and v_n^{im} into the system Equations (21) and (22), and taking into account Equation (18), the solution for the optimal normalized input voltages $v_n^{opt} = v_n^{re,opt} + jv_n^{im,opt}$ at each input port is found.

$$v_n^{re,opt} = 0, \tag{23}$$

$$v_n^{im,opt} = \frac{k_{0n} Q_n (\eta_{max} - 1)}{2\eta_{max}} v_0. \tag{24}$$

If the values of the normalized input port voltages v_n ($n = 1, \ldots, N$) equal Equations (23) and (24), the maximum attainable efficiency η_{max} is reached. It is important to note that the voltages here are not only expressed as a function of the parameters of the circuit network (which are known and fixed) and the reference output voltage v_0, but also of the maximum efficiency η_{max}, which is (for now) an unknown value. In Section 4.3, the value of η_{max} will be determined.

4.2. Optimal Input and Output Power

By substituting Equations (23) and (24) into Equations (15) and (17), the input power P_{in}^{opt} and output power P_{out}^{opt} at the maximum efficiency configuration are determined:

$$P_{in}^{opt} = \frac{v_0^2(1-\eta_{max}^2)\alpha_N^2}{8Q_0\eta_{max}^2}, \tag{25}$$

$$P_{out}^{opt} = -\frac{v_0^2}{2Q_0}\left[1 + \frac{(\eta_{max}-1)\alpha_N^2}{2\eta_{max}}\right], \tag{26}$$

where the following notation is introduced

$$\alpha_N^2 = \sum_{n=1}^{N}\alpha_n^2, \tag{27}$$

with

$$\alpha_n = k_{0n}\sqrt{Q_0 Q_n}. \tag{28}$$

The parameter α_n is named *the extended kQ-product* of the link between the n-th transmitter and the receiver, analogous to [27–29]. The variable α_N is called the *system kQ-product*, a naming borrowed from [14,15,29]. The introduction of these variables seems artificial at this point, but will be further discussed in Section 5.

The value of η_{max} is still unknown and will be determined in the following subsection.

4.3. Maximum Efficiency

A quadratic equation in η_{max} is found by substituting Equations (25) and (26) in Equation (18):

$$\eta_{max}^2 - \left(2 + \frac{4}{\alpha_N^2}\right)\eta_{max} + 1 = 0. \tag{29}$$

In order to alleviate the notation, the symbol γ is introduced:

$$\gamma = \sqrt{1+\alpha_N^2}. \tag{30}$$

The quadratic Equation (29) results in two solutions:

$$\eta_{max,1} = \frac{\gamma-1}{\gamma+1}, \tag{31}$$

and

$$\eta_{max,2} = \frac{\gamma+1}{\gamma-1}. \tag{32}$$

Equation (32) is physically not possible since $0 \leq \eta_{max} \leq 1$. The maximum attainable power-transfer efficiency η_{max} is therefore expressed by Equation (31).

The maximum efficiency η_{max} is now determined as a function of the characteristics of the circuit parameters only, which implies that the optimal input voltages Equations (23) and (24), input Equation (25), and also output power Equation (26) are expressed as functions of the circuit characteristics only.

For example, the optimal output and input power are given by

$$P_{out}^{opt} = \frac{\gamma}{2Q_0}v_0^2, \tag{33}$$

$$P_{in}^{opt} = \frac{P_{out}^{opt}}{\eta_{max}} = \frac{\gamma}{2Q_0} \frac{\gamma+1}{\gamma-1} v_0^2. \tag{34}$$

4.4. Optimal Input Voltages, Currents, and Admittances

Combining Equations (23), (24), and (31), the optimal normalized input voltages $v_n^{opt} = v_n^{re,opt} + jv_n^{im,opt}$ are found:

$$v_n^{re,opt} = 0, \tag{35}$$

$$v_n^{im,opt} = \frac{k_{0n}Q_n}{1-\gamma} v_0. \tag{36}$$

From Equations (11), the optimal normalized input currents $i_n^{opt} = i_n^{re,opt} + ji_n^{im,opt}$ follow

$$i_n^{re,opt} = \frac{v_0}{1-\gamma} \sum_{i=1}^{N} k_{0i} k_{in} Q_i, \tag{37}$$

$$i_n^{im,opt} = \frac{\gamma k_{0n}}{1-\gamma} v_0. \tag{38}$$

The optimal normalized input admittance $y_n^{in,opt}$ at port n thus equals

$$y_n^{in,opt} = \frac{i_n^{opt}}{v_n^{opt}} = \frac{\gamma}{Q_n} - j\frac{1}{k_{0n}Q_n} \sum_{i=1}^{N} k_{0i} k_{in} Q_i. \tag{39}$$

From this equation, it can be concluded that the cross-coupling between the transmitters can be compensated by a normalized shunt inductance b_n^S equal to

$$b_n^S = \frac{1}{k_{0n}Q_{nn}} \sum_{i=1}^{N} k_{0i} k_{in} Q_{ii}, \tag{40}$$

or unnormalized

$$B_n^S = \frac{g_{nn}}{b_{0n}} \sum_{i=1}^{N} \frac{b_{0i} b_{in}}{g_{ii}}. \tag{41}$$

Equation (39) implies that the optimal input conditions can be obtained by a set of N independent current generators operating in maximum power-transfer conditions. The internal normalized shunt admittances of these generators are

$$y_n^S = (y_n^{in,opt})^* = \frac{\gamma}{Q_n} + j\frac{1}{k_{0n}Q_n} \sum_{i=1}^{N} k_{0i} k_{in} Q_i, \tag{42}$$

and their normalized currents are

$$i_n^S = i_n^{opt} + y_n^S v_n^{opt} = 2jk_{0n} \frac{\gamma}{1-\gamma} v_0. \tag{43}$$

In this way, maximum power transfer from the generators to the network is achieved. At this point, the maximum-efficiency solution also becomes the one that provides the maximum output power.

The corresponding unnormalized values of the shunt admittances and currents of these generators are (Figure 2a)

$$Y_n^S = \gamma g_{nn} + j\frac{g_{nn}}{b_{0n}} \sum_{i=1}^{N} \frac{b_{0i} b_{in}}{g_{ii}}, \tag{44}$$

$$I_n^S = 2jb_{0n} \frac{\gamma}{1-\gamma} V_0. \tag{45}$$

Since v_0 was chosen as reference phasor, the condition for the input current sources can be practically achieved by imposing that the ratios of the input currents must satisfy

$$\frac{i_G^S}{k_{01}} = \frac{i_2^S}{k_{02}} = \ldots = \frac{i_N^S}{k_{0N}} \tag{46}$$

or

$$\frac{I_1^S}{b_{01}} = \frac{I_2^S}{b_{02}} = \ldots = \frac{I_N^S}{b_{0N}}. \tag{47}$$

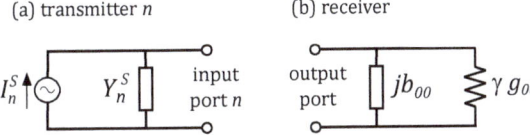

Figure 2. Maximum-efficiency solution for (**a**) the transmitter ports with generator I_n^S and internal shunt admittance Y_n^S, and (**b**) the receiver port with the optimal load susceptance and load conductance.

4.5. Optimal Load Admittance

Finally, it is possible to determine the optimal load that realizes the maximum-efficiency solution. The optimal normalized load admittance is given by

$$y_0^{L,opt} = g_0^{L,opt} + jb_0^{L,opt} = -\frac{i_0^{opt}}{v_0^{opt}}, \tag{48}$$

where $g_0^{L,opt}$ and $b_0^{L,opt}$ are the optimal normalized load conductance and susceptance, respectively.

From Equation (11), it follows that

$$i_0^{re,opt} = \frac{v_0}{Q_0} + \sum_{n=1}^{N} k_{0n} v_n^{im,opt}, \tag{49}$$

$$i_0^{im,opt} = -k_{00} v_0 - \sum_{n=1}^{N} k_{0n} v_n^{re,opt}. \tag{50}$$

Substituting Equations (35) and (36) results into

$$i_0^{re,opt} = -\frac{\gamma}{Q_0} v_0, \tag{51}$$

$$i_0^{im,opt} = -k_{00} v_0, \tag{52}$$

which leads to the optimal normalized load conductance $g_0^{L,opt}$ and load susceptance $b_0^{L,opt}$ as functions of the parameters of the network:

$$g_0^{L,opt} = \frac{\gamma}{Q_0}, \tag{53}$$

$$b_0^{L,opt} = k_{00}. \tag{54}$$

The corresponding unnormalized values are (Figure 2b)

$$G_0^{L,opt} = \gamma g_{00}, \tag{55}$$

$$B_0^{L,opt} = b_{00}. \tag{56}$$

5. Discussion

In order to practically achieve the maximum attainable efficiency η_{max}, three conditions must be met simultaneously:

1. The value of the input current sources must satisfy Equation (47) (Figure 2a).
2. Shunt susceptances with value found in Equation (41) must be connected to each input port (Figure 2a).
3. The output load conductance and susceptance must equal Equations (55) and (56), respectively (Figure 2b).

Additionally, if the internal shunt admittances of the generators equal Equation (44), the maximum-efficiency solution also becomes the one that maximizes the output power.

The first condition indicates that, the higher the coupling between transmitter n and the receiver, the lower the necessary value of the current source for that transmitter. From Equations (35), (36), and (45), it follows that the phasors of the optimal current sources and optimal voltages of all input ports are orthogonal to the reference output voltage V_0.

Instead of applying current sources at each transmitter, one could also apply voltage sources for which the ratios must satisfy

$$V_1 : V_2 : \ldots : V_n = \frac{b_{01}}{g_{11}} : \frac{b_{02}}{g_{22}} : \ldots : \frac{b_{0n}}{g_{nn}}. \tag{57}$$

The optimal input voltages are in other words determined by the ratios between the transmitter–receiver coupling strength and the resistive losses of the transmitter.

The second condition, the insertion of shunt susceptances at the input port, is necessary to compensate for the cross-coupling between the transmitters. If no cross-coupling is present, no shunt susceptances have to be inserted, as can be seen by Equation (41).

Under the optimal conditions, the maximum efficiency η_{max} given by Equation (31) is reached. Notice that η_{max} is independent on the cross-coupling between the transmitters; the optimization of input voltages V_n^{opt} and load admittance $Y_0^{L,opt}$ eliminates the influence of the cross-coupling. Nevertheless, the presence of the shunt susceptances at the input ports is necessary for achieving η_{max} and to ensure that the optimal voltages V_n^{opt} are reached from the current sources I_n^S that supply the power for the CPT system.

The third condition refers to the terminating load. The optimal load conductance $G_0^{L,opt}$ of the receiver is proportionate to its parasitic conductance g_{00}. For high coupling ($\alpha_N \gg 1$) between the transmitters and receiver, the optimal conductance can be approximated by $G_0^{L,opt} = g_{00}\alpha_N$. The output load susceptance must equal Equation (56) and thus cancels out the self-susceptance of the receiver resonator.

The optimal terminating load admittance corresponds to the value found in scientific literature for a CPT system with a single transmitter ($N = 1$) coupled to a single receiver [30–32].

Notice that not only the maximum efficiency η_{max}, but also the optimal load and input current ratios are independent on the cross-coupling between the transmitters. For an uncoupled system, the maximum efficiency η_{max} and optimal load are the same as for a coupled system, since the shunt susceptances at the input ports compensate for the cross-coupling. This does not imply that the *efficiency* is not influenced by cross-coupling for a *general* CPT system; it is the *maximum* efficiency that is invariant for cross-coupling for an *optimized* system towards efficiency.

The efficiency rises with higher couplings between transmitter and receiver, and lower conductances $g_{00}, g_{11}, \ldots, g_{NN}$ of the system.

It is not surprising that the maximum efficiency η_{max} is expressed as a function of a single variable; it is a general property of *any* reciprocal power transfer system that the efficiency can be stated as a function of a single scalar [28]. In the context of WPT with multiple transmitters and/or receivers, this variable is often called the system kQ-product [14,15,20,23]

From Equation (27), it can be concluded that the square of the system kQ-product equals the sum of the squares of the kQ products of each individual transmitter–receiver link. Determining the kQ product for each single transmitter–receiver pair thus results in a prediction for the entire system's power-transfer efficiency.

The higher the system kQ-product α_N, the higher the efficiency of the system, as can be seen by Equations (27), (28), and (31). The maximum efficiency η_{max} of the CPT system can thus be increased by adding more transmitters to the system, even if the transmitters themselves are coupled. Indeed, the cross-coupling between the transmitters can be compensated by the shunt susceptances B_n^S at the input ports. There is no optimal number of transmitters. The more transmitters, the higher the maximum attainable efficiency η_{max}.

6. Numerical Verification

In order to validate the theory, an example of CPT system with three transmitters ($N = 3$) and a single receiver is considered; it is assumed that there is a cross-coupling present between the transmitters themselves (Figure 3a). The parameters within the dashed rectangle are assumed to be given and fixed, including the coupling strengths. They can be represented by the admittance matrix Y. In order to optimize the CPT system towards efficiency, it is possible to act on the value of the load $Y_0^L = G_0^L + jB_0^L$, the supply current sources I_G^S, and the input shunt susceptances B_n^S.

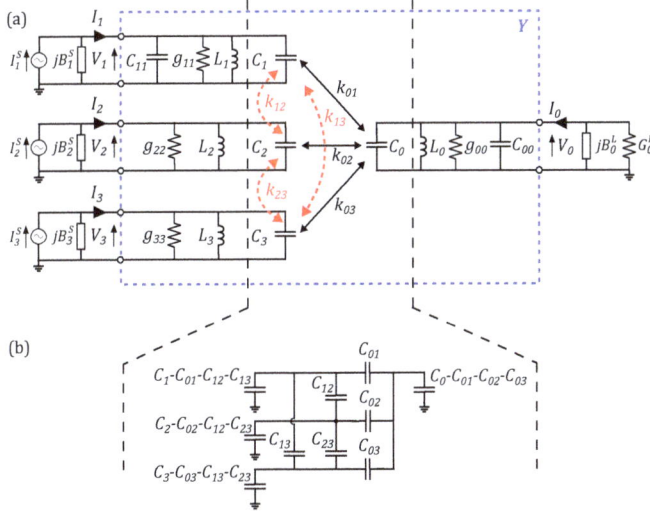

Figure 3. (a) Equivalent circuit of a capacitive wireless power transfer system with 3 transmitters (left) and a single receiver (right). The (desired) electric couplings between transmitters and receiver are depicted by the full arrows. The (undesired) cross-couplings between transmitters themselves are indicated by the dashed arrows. (b) Applied equivalent circuit for the simulation of the capacitive coupling.

The numerical values indicated in Table 1 are considered, the operating frequency is f_0=10 MHz. No specific design consideration is assumed; a range of different desired and undesired coupling factors (Table 2) were chosen to verify the analytical derivation. Further, it is assumed that the receiver and the first transmitter have a self-susceptance C_{00} and C_{11}, respectively.

Table 1. Given network simulation parameters for the analyzed numerical example.

Quantity	Value	Quantity	Value
f_0	10.0 MHz	I_1^S	100 mA
g_{00}	1.00 mS	C_0	350 pF
g_{11}	1.00 mS	C_1	350 pF
g_{22}	1.50 mS	C_2	300 pF
g_{33}	0.50 mS	C_3	275 pF
C_{00}	500 pF	C_{11}	500 pF

Table 2. Coupling factors of the analyzed example.

Desired Couplings	Value	Undesired Couplings	Value
k_{01}	14.3%	k_{12}	6.2%
k_{02}	46.3%	k_{13}	16.1%
k_{03}	32.2%	k_{23}	3.5%

Electric coupling is realized by the coupled capacitors C_j ($j = 0, 1, 2, 3$). In each transmitter and in the receiver, a resonant circuit is constructed by adding a shunt inductor L_j with value

$$L_j = \frac{1}{\omega_0^2 C_j}. \tag{58}$$

The corresponding values are given in Table 3.

In order to verify the analytical formulas, the numerical example has been simulated in AWR NI. Figure 3b depicts the applied equivalent circuit for the simulation of the capacitive coupling [25].

First of all, the admittance matrix of the link has been calculated, obtaining the following values:

$$Y = \begin{bmatrix} 1+31.42j & -1.26j & -3.14j & -3.14j \\ -1.26j & 1.5 & -0.628j & -9.42j \\ -3.14j & -0.628j & 0.5 & -6.28j \\ -3.14j & -9.42j & -6.28j & 1+31.42j \end{bmatrix} \cdot 10^{-3}. \tag{59}$$

By using the values reported in Equation (59), the extended kQ-product α_n, the system kQ-product α_N, and the inductors L_j, can be calculated from Equations (28), (27), and (58), respectively, and are listed in Table 3.

Table 3. Calculated network simulation parameters for the example capacitive power transfer (CPT) system.

Quantity	Value	Quantity	Value
L_0	724 nH	α_1	3.1
L_1	724 nH	α_2	7.7
L_2	844 nH	α_3	8.9
L_3	921 nH	α_N	12.2

As per the maximum efficiency, a value of 84.9% is attainable, according to Equation (31). The parameters at which the maximum efficiency configuration is reached are listed in Table 4. A current source I_1^S of the first transmitter of 100 mA was chosen, resulting in the optimal current sources of the other transmitters, according to Equation (47).

The optimal load admittance is calculated by Equations (55) and (56). The optimal load susceptance is negative, i.e., it corresponds to a shunt inductor $L_0^{L,opt}$.

Additionally, according to Equation (41) a shunt susceptance at each transmitter side is necessary to compensate the transmitter's cross-coupling. At the first transmitter, the shunt susceptance is an inductor L_1^S. At the second and third transmitter, it is found that the shunt susceptances are capacitors C_2^S and C_3^S. All the values calculated from theoretical formulas are summarized in Table 4.

Table 4. Calculated values for the maximum efficiency solution.

Quantity	Value	Quantity	Value
I_2^S	300 mA	I_3^S	200 mA
$G_0^{L,opt}$ ($R_0^{L,opt}$)	12.2 mS (81.9 Ω)	L_1^S	974 nH
$L_0^{L,opt}$	506 nH	C_2^S	30.0 pF
η_{max}	84.9%	C_3^S	17.5 pF

First, the simulation is executed at the maximum-efficiency configuration of Table 4. The simulation program returns a power-transfer efficiency η of 84.9% and confirms that the output voltage V_0 is orthogonal to the input voltages and currents.

Next, the load conductance and load susceptance are varied, respectively, while keeping the other parameters fixed at their optimal value given in Table 4. The simulation results are depicted in Figures 4 and 5.

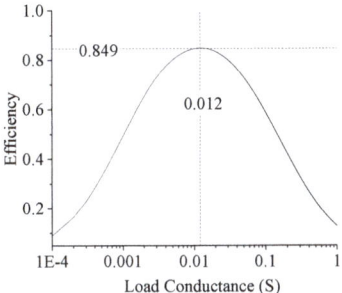

Figure 4. The simulated efficiency η as a function of varying load conductance for the given system of three transmitters and a single receiver. The load conductance is varied, while keeping the other system parameters at their optimal value.

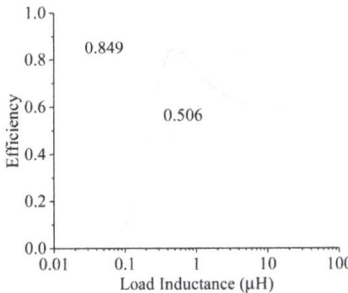

Figure 5. The simulated efficiency η as a function of varying shunt load inductance for the given system of three transmitters and a single receiver. The load inductance is varied, while keeping the other system parameters at their optimal value.

Regarding effect on the efficiency of the compensating shunt susceptances L_1^S, C_2^S, and C_3^S, this is investigated in Figures 6 and 7. The simulated efficiency confirms that maximum efficiency is obtained by using the compensating shunt susceptances calculated by using the theoretical formulas.

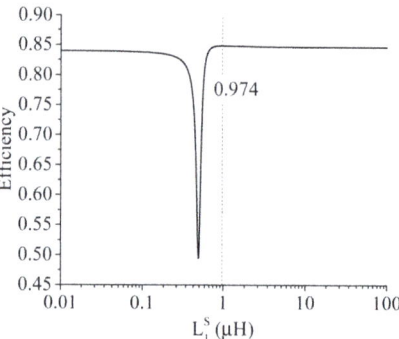

Figure 6. The simulated efficiency η as a function of the compensating susceptance L_1^S. L_1^S is varied, while keeping the other system parameters at their optimal value.

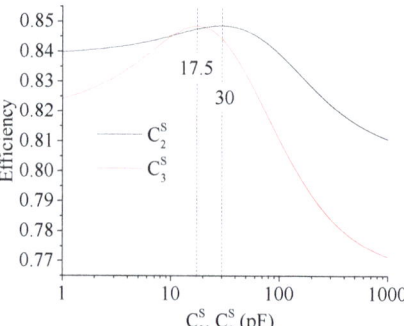

Figure 7. The simulated efficiency η as a function of the compensating susceptance C_2^S and C_3^S. C_2^S (black curve) or C_3^S (red curve) is varied, while keeping the other system parameters at their optimal value.

Additionally, the simulation confirms that the shunt susceptances at the input ports eliminate the cross-coupling. For the circuit without the compensating shunt susceptances and no cross-coupling, the same maximum efficiency of 84.9% is reached.

Finally, the effect of the amplitude of the input currents has been analyzed. The achieved results are summarized in Figure 8. It is observed that the efficiency is always lower than the optimal current distribution from Table 4 (i.e., $I_2/I_1 = 3$ and $I_3/I_1 = 2$). For example, if all input current sources are equal to 100 mA, an efficiency of 76.2% is attained.

The case of voltage sources has been also analyzed; in this case, by using Equation (57) and the values of Equation (59), the optimal voltage ratios can be determined, e.g.,

$$\frac{V_2}{V_1} = \frac{\frac{b_{02}}{g_{22}}}{\frac{b_{01}}{g_{11}}} = 2. \tag{60}$$

Analogously, the optimal ration $V_3/V_1 = 4$ is found. This is confirmed by circuital simulation results summarized in Figure 9.

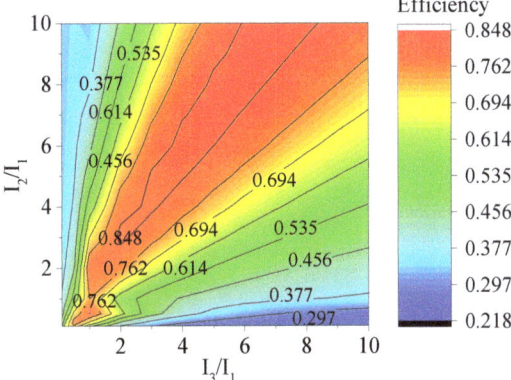

Figure 8. The simulated efficiency η as a function of the ratio of the input currents. The efficiency is always lower than the optimal current distribution $I_2/I_1 = 3$ and $I_3/I_1 = 2$. For example, if all input current sources are equal (i.e., $I_2/I_1 = 1$ and $I_3/I_1 = 1$), an efficiency of 76.2% is attained.

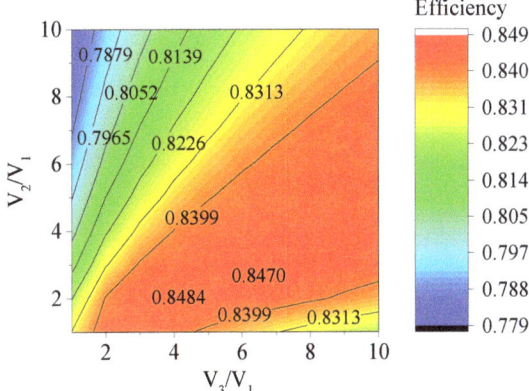

Figure 9. The simulated efficiency η as a function of the ratio of the input voltages.

In conclusion, circuital simulations confirm the data provided by the theory for the analyzed example: a maximum power-transfer efficiency is achieved for the optimal values of Table 4, calculated according to analytical derivation.

7. Conclusions

A general CPT system with any number of transmitters and a single receiver was optimized towards power-transfer efficiency. It was shown that in order to maximize the efficiency of a system with given wireless links and couplings, three conditions must be fulfilled simultaneously. First, the ratio of the input current sources is dependent on the coupling between each transmitter and the receiver, given by Equation (47). Secondly, the undesired cross-coupling between the transmitters themselves can be eliminated by adding appropriate shunt susceptances, given by Equation (41), at the input terminals. Finally, the optimal load is purely resistive, equal to Equation (55), if the receiver has no self-susceptance. If a self-susceptance is present at the receiver's side, a compensating load susceptance is required.

Additionally, by conjugate-matching the internal shunt admittance of the generators, the maximum-efficiency solution coincides with the configuration that maximizes the output power.

It was shown that the maximum achievable efficiency η_{max}, the optimal loads, and the optimal input currents are independent on the cross-coupling between the transmitters, since this unwanted cross-coupling can be entirely annihilated with the transmitter shunt susceptances B_n^S. As a result, it is possible to increase the system efficiency by adding more transmitters, and compensating every time for transmitter cross-coupling.

The expression for the extended kQ-factor for each transmitter–receiver link was determined, allowing an estimate of the maximum efficiency of the CPT system via the system kQ-product.

Finally, the analytical derivation was verified by simulation of an example CPT system with three transmitters and a single receiver. Measurements on a CPT setup with multiple transmitters are required to confirm the accuracy of the analytical results and are part of future research.

Author Contributions: Conceptualization, methodology, B.M., M.M.; validation, B.M., G.M., M.M.; writing–original draft preparation, B.M.; writing–review and editing, A.C., G.M., M.M. All authors have read and agreed to the published version of the manuscript.

Funding: This research received no external funding.

Acknowledgments: The authors would like to remember the colleague Franco Mastri who suddenly passed away on April 3rd, 2020. He was a great colleague and a profound scientist. Fundamental discussions and studies on the theoretical modelling of near-field WPT systems were of great inspiration also for the results presented in this work.

Conflicts of Interest: The authors declare no conflict of interest.

References

1. Zhu, Q.; Zang, S.; Zou, L.J.; Zhang, G.; Su, M.; Hu, A.P. Study of coupling configurations of capacitive power transfer system with four metal plates. *Wireless Power Transfer* **2019**, *6*, 97–112. [CrossRef]
2. Wang, K.; Sanders, S. Contactless USB—A capacitive power and bidirectional data transfer system. In Proceedings of the IEEE Applied Power Electronics Conference and Exposition-APEC, Fort Worth, TX, USA, 16–20 March 2014; pp. 1342–1347.
3. Hu, A.P.; Liu, C.; Li, H.L. A novel contactless battery charging system for soccer playing robot. In Proceedings of the IEEE 15th International Conference on Mechatronics and Machine Vision in Practice, Auckland, New Zealand, 2–4 December 2008; pp. 646–650.
4. Culurciello, E.; Andreou, A.G. Capacitive inter-chip data and power transfer for 3-D VLSI. *IEEE Trans. Circ. Syst. II Express Briefs* **2006**, *53*, 1348–1352. [CrossRef]
5. Mostafa, T.M.; Muharam, A.; Hattori, R. Wireless battery charging system for drones via capacitive power transfer. In Proceedings of the IEEE PELS Workshop on Emerging Technologies: Wireless Power Transfer (WoW), Chongqing, China, 20–22 May 2017; pp. 1–6.
6. Sodagar, A.M.; Amiri, P. Capacitive coupling for power and data telemetry to implantable biomedical microsystems. In Proceedings of the IEEE 4th International IEEE/EMBS Conference on Neural Engineering, Antalya, Turkey, 29 April–2 May 2009; pp. 411–414.
7. Jegadeesan, R.; Agarwal, K.; Guo, Y.X.; Yen, S.C.; Thakor, N.V. Wireless power delivery to flexible subcutaneous implants using capacitive coupling. *IEEE Trans. Microwave Theory Tech.* **2016**, *65*, 280–292. [CrossRef]
8. Ramos, I.; Afridi, K.; Estrada, J.A.; Popović, Z. Near-field capacitive wireless power transfer array with external field cancellation. In Proceedings of the IEEE Wireless Power Transfer Conference (WPTC), Aveiro, Portugal, 5–6 May 2016; pp. 1–4.
9. Miyazaki, M.; Abe, S.; Suzuki, Y.; Sakai, N.; Ohira, T.; Sugino, M. Sandwiched parallel plate capacitive coupler for wireless power transfer tolerant of electrode displacement. In Proceedings of the IEEE MTT-S International Conference on Microwaves for Intelligent Mobility (ICMIM), Nagoya, Japan, 19–21 March 2017; pp. 29–32.
10. Dai, J.; Ludois, D.C. Wireless electric vehicle charging via capacitive power transfer through a conformal bumper. In IEEE Applied Power Electronics Conference and Exposition (APEC), Charlotte, NC, USA, 15 March 2015; pp. 3307–3313.

11. Sakai, N.; Itokazu, D.; Suzuki, Y.; Sakihara, S.; Ohira, T. One-kilowatt capacitive Power Transfer via wheels of a compact Electric Vehicle. In Proceedings of the IEEE Wireless Power Transfer Conference (WPTC), Aveiro, Portugal, 5–6 May 2016; pp. 1–3.
12. Sinha, S.; Kumar, A.; Regensburger, B.; Afridi, K.K. A new design approach to mitigating the effect of parasitics in capacitive wireless power transfer systems for electric vehicle charging. *IEEE Trans. Transport. Electr.* **2019**, 1040–1059. [CrossRef]
13. Monti, G.; Che, W.; Wang, Q.; Costanzo, A.; Dionigi, M.; Mastri, F.; Mongiardo, M.; Perfetti, R.; Tarricone, L.; Chang, Y. Wireless power transfer with three-ports networks: Optimal analytical solutions. *IEEE Trans. Circ. Syst. I Regul. Pap.* **2016**, *64*, 494–503. [CrossRef]
14. Ujihara, T.; Duong, Q.T.; Okada, M. kQ-product analysis of inductive power transfer system with two transmitters and two receivers. In Proceedings of the IEEE Wireless Power Transfer Conference, Taipei, Taiwan, 10–12 May 2017.
15. Duong, Q.T.; Okada, M. kQ-product formula for multiple-transmitter inductive power transfer system. *IEICE Electron. Express* **2017**, *14*, 20161167. [CrossRef]
16. Arakawa, T.; Goguri, S.; Krogmeier, J.V.; Kruger, A.; Love, D.J.; Mudumbai, R.; Swabey, M.A. Optimizing wireless power transfer from multiple transmit coils. *IEEE Access.* **2018**, *6*, 23828–23838. [CrossRef]
17. Lang, H.D.; Sarris, C.D. Semidefinite relaxation-based optimization of multiple-input wireless power transfer systems. *IEEE Trans. Microwave Theory Tech.* **2017**, *65*, 4294–4306. [CrossRef]
18. Lang, H.D.; Ludwig, A.; Sarris, C.D. Convex optimization of wireless power transfer systems with multiple transmitters. *IEEE Trans. Antennas Propagat.* **2014**, *62*, 4623–4636. [CrossRef]
19. Monti, G.; Wang, Q.; Che, W.; Costanzo, A.; Mastri, F.; Mongiardo, M. Maximum wireless power transfer for multiple transmitters and receivers. In Proceedings of the IEEE MTT-S International Conference on Numerical Electromagnetic and Multiphysics Modeling and Optimization (NEMO), Beijing, China, 27–29 July 2016; pp. 1–3.
20. Duong, Q.T.; Okada, M. Maximum efficiency formulation for inductive power transfer with multiple receivers. *IEICE Electron. Express* **2016**, *13*, 20160915–20160915. [CrossRef]
21. Yoon, I.J.; Ling, H. Investigation of near-field wireless power transfer under multiple transmitters. *IEEE Antennas Wirel. Propag. Lett.* **2011**, *10*, 662–665. [CrossRef]
22. Minnaert, B.; Stevens, N. Optimal analytical solution for a capacitive wireless power transfer system with one transmitter and two receivers. *Energies* **2017**, *10*, 1444. [CrossRef]
23. Minnaert, B.; Mongiardo, M.; Costanzo, A.; Mastri, F. Maximum Efficiency Solution for Capacitive Wireless Power Transfer with N Receivers. *Wireless Power Transfer* **2020**, 1–11. [CrossRef]
24. Huang, L.; Hu, A.P. Defining the mutual coupling of capacitive power transfer for wireless power transfer. *Electron. Lett.* **2015**, *51*, 1806–1807. [CrossRef]
25. Hong, J.S.G.; Lancaster, M.J. *Microstrip Filters for RF/Microwave Applications*, 1st ed.; John Wiley & Sons: New York, NY, USA, 2001; pp. 235–253.
26. Boyd, S.; Vandenberghe, L. *Convex Optimization*, 2nd ed.; Cambridge University Press: Cambridge, UK, 2004; pp. 140.
27. Ohira T. Extended k-Q product formulas for capacitive-and inductive-coupling wireless power transfer schemes. *IEICE Electron. Express* **2014**, *11*, 20140147. [CrossRef]
28. Minnaert, B.; Stevens, N. Single variable expressions for the efficiency of a reciprocal power transfer system. *Int. J. Circ. Theory Appl.* **2017**, *10*, 1418–1430. [CrossRef]
29. Sugiyama, R.; Duong, Q.T.; Okada, M. kQ-product analysis of multiple-receiver inductive power transfer with cross-coupling. In Proceedings ot the International Workshop on Antenna Technology: Small Antennas, Innovative Structures, and Applications (iWAT), Athens, Greece, 1–3 March 2017.
30. Dionigi, M.; Mongiardo, M.; Monti, G.; Perfetti, R. Modelling of wireless power transfer links based on capacitive coupling. *Int. J. Numer. Model. Electron. Netw. Dev. Fields* **2017**, *30*, e2187. [CrossRef]

31. Kracek, J.; Svanda, M. Analysis of Capacitive Wireless Power Transfer. *IEEE Access.* **2018**, *7*, 26678–26683. [CrossRef]
32. Kim, D.H.; Ahn, D. Optimization of Capacitive Wireless Power Transfer System for Maximum Efficiency. *J. Electr. Eng. Technol.* **2020**, *15*, 343–352. [CrossRef]

© 2020 by the authors. Licensee MDPI, Basel, Switzerland. This article is an open access article distributed under the terms and conditions of the Creative Commons Attribution (CC BY) license (http://creativecommons.org/licenses/by/4.0/).

Article

Scaling-Factor and Design Guidelines for Shielded-Capacitive Power Transfer

Aam Muharam [1,2], Suziana Ahmad [1,3] and Reiji Hattori [1,*]

1. Interdisciplinary Graduate School of Engineering Sciences, Kyushu University, Fukuoka 816-8580, Japan; aam.muharam@lipi.go.id (A.M.); binti.ahmad.suziana.749@s.kyushu-u.ac.jp (S.A.)
2. Research Centre for Electrical Power and Mechatronics, Indonesian Institute of Sciences, Bandung 40135, Indonesia
3. Faculty of Electrical and Electronic Engineering Technology, Universiti Teknikal Malaysia Melaka, Melaka 76100, Malaysia
* Correspondence: hattori@gic.kyushu-u.ac.jp; Tel.: +81-92-583-7887

Received: 11 July 2020; Accepted: 13 August 2020; Published: 16 August 2020

Abstract: This paper introduces scaling-factor and design guidelines for shielded-capacitive power transfer (shielded-CPT) systems, offering a simplified design process, coupling-structure optimization, and consideration of safety. A novel scaling-factor-analysis method is proposed by determining the configuration of the coupling structure that improves system safety and increases operating efficiency while minimizing the gap between the shield and the coupler plate. The inductor-series resistance is also analyzed to study the loss efficiency in the shielded-CPT system. The relationship among the shield-coupler gap, distance between the couplers, conductive-plate size, and delivered power is examined and presented. The proposed method is validated by implementing the shielded-CPT system with hardware and the result suggests that the proposed method can be used to design shielded-CPT systems with scaling-factor and safety considerations.

Keywords: capacitive wireless power transfer; wireless power transmission; electric field; shielded-capacitive power transfer; design guidelines

1. Introduction

Capacitive power transfer (CPT) is an alternative approach to wireless power transfer (WPT). Rather than using a magnetic field, CPT uses a quasi-static electric field (EF) to deliver power from the primary side to the secondary side through a capacitor formed by electrodes belonging to physically separate devices [1–3]. Murata Electronics Europe adopted this method, and it has become popular because of good galvanic isolation, low cost, and the potential for operation at a higher frequency rating than that of the magnetic core [4,5]. The CPT system has been widely used in previous applications, such as electric vehicle (EV) charging [6–10], drones [11–13], variable message displays [14], and others. Previous research on low-power [11,15] and high-power [16] applications of CPT has been conducted with a reported efficiency of more than 90%. The CPT system offers advantages in lightweight, contactless, and electromagnetic interference (EMI) reduction. These advantages allow the CPT system to compromise a suitability-integrated system for available EV charging, such as in-vehicle grid interaction [17,18] and the grid-tied plug-in EV charging system [19,20]. In the study by [17,19], the potential of the CPT system can be used to replace the cable between the grid-connected EV Supply Equipment (EVSE)-Plug-in EV (PEV) (EVSE-PEV) and the EV itself.

Furthermore, much research has been conducted on reducing EF emissions using techniques, such as the single-wire system [14,21,22] and the six-plate coupling-interface method [23]. In the study by [23], the coupler has a thickness of 1.9 cm and a gap of 15 cm. The efficiency of the system was reported as 91.6% when delivering power of 1.97 kW, whereas the safety range of EF emissions was

>0.4 m from the coupler. However, the method of calculating the resonant and component parameters was too complex. The concept of shielded-CPT was introduced in EV charging applications [24,25], using two extra plates to cover the coupler on each side.

This paper proposes a deep analysis of scaling-factor and design guidelines to achieve a compact shielded-CPT system that meets design requirements and safety considerations. The contributions of this paper are as follows.

- A novel method for analyzing the coupling interface of a shielded-CPT system is introduced. This method allows determination of the configuration of coupling structures intended for overall improved safety and higher operating efficiency in CPT systems while being safe for human use and allowing the possibility of thinner modules.
- A design guideline is introduced for scaling and optimizing the shielded-CPT system such that requirements, specific conditions, and safety level standards are met.

2. Scaling Model and Analysis

The proposed shielded-CPT structure is constructed as a conventional CPT coupling-plate interface with two additional plates behind each side. The coupling structure builds a six-plate CPT system consisting of a power transfer part and a shielding part as seen in Figure 1. With this configuration, the circuit parameter is optimized to the required power and efficiency, the size and distance of the coupling in consideration to the safety level of air breakdown voltage and the stray of EF. By introducing the extra plates, the EF-emission characteristic was observed through field-simulation and hardware experiments [23,24]. In these studies, the six-plate CPT system shows that the EF emission has been reduced significantly compared to the four-plate systems [26].

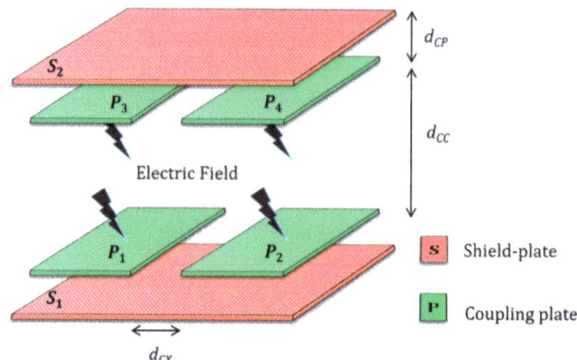

Figure 1. Structure of shielded-capacitive power transfer (CPT) coupling-interface system.

The circuit model of the complete shielded-CPT system is shown in Figure 2. Four parts of the circuit model comprise a WPT system with a capacitive coupling interface. A switch network can be implemented by a single-ended Class-E power amplifier, half-bridge, or full-bridge inverter system [27–30]. A 50 Ω coaxial cable is used in the proposed system, and a balanced-to-unbalanced (Balun) transformer is coupled to the resonant inductors, providing a balanced condition of the voltage waveform and a stable ground reference to the coupling system [31]. For simplicity of modeling, these two parts will be omitted in the scaling-factor analysis.

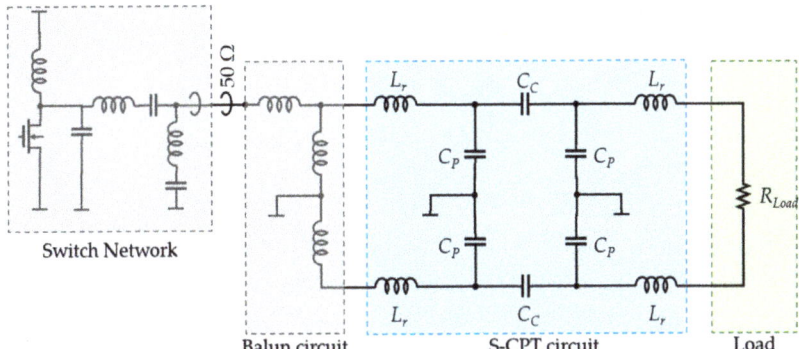

Figure 2. Circuit model of the complete shielded-CPT system.

2.1. Circuit Model Analysis of the Shielded-CPT System

Figure 3 presents the analysis of the coupling interface of the proposed shielded-CPT system. The input-voltage source, V_0, produces a sinewave alternate current (AC) voltage that is applied to the input terminal of the shielded-CPT circuit. Accordingly, the resonant frequency is tuned by the value of the series-resonant inductor, L_r, coupling capacitance, C_C, and parasitic capacitance, C_P, on the primary and secondary sides. The load resistance, R_{Load}, is connected through the resonant inductors in the secondary side. As an assumption, the circuit topology involves of symmetry parameters for both placement and size.

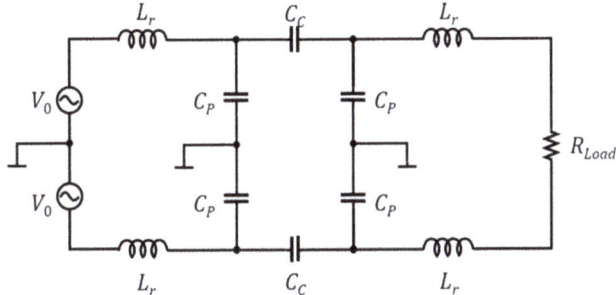

Figure 3. Circuit model for S-CPT-coupling-capacitance analysis.

The shielded-CPT circuit can be analyzed through three approximations. First, the primary (transmitter)-side impedance can be calculated under the assumption that all components on the primary side are modeled by a single parallel resistance, R_L, if a resonant condition occurs between them. Next, the secondary (receiver)-side impedance was assumed to be in a resonant condition with the receiver side; thus, the circuit is modeled by a single parallel resistance, R_L. The last approximation is a combination of the previous two, namely, the primary and secondary impedances match. Secondary-side-impedance-matched analysis is used for this paper. Figure 4 illustrates the simplified circuit model for the process of analyzing the coupling capacitance, C_C. It consists of a single input-voltage source, V_0, with frequency f, connected to the series-resonant inductor, L, with equivalent series resistance (ESR), R_S. The circuit is coupled with the parasitic capacitance, C_P, in parallel with the coupling capacitance, C_C. here we assume that the secondary side has a resonant condition. The load resistance R_{Load} and the L-matching circuit can be simplified as the load, R_L.

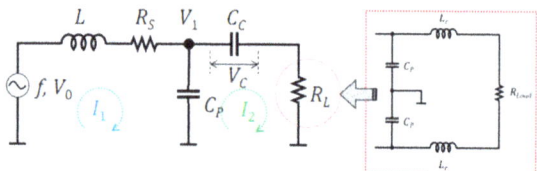

Figure 4. Simplified circuit model for system analysis.

Using the circuit diagram in Figure 4, Equation (1) is obtained based on the sinusoidal approximation for the I_1 current loop with angular frequency, ω, using Kirchhoff laws (KCL):

$$V_0 = (j\omega L + R_S)I_1 + \frac{1}{j\omega C_p}(I_1 - I_2) \tag{1}$$

In addition, applying KCL to the I_2 loop yields the relationship:

$$0 = -\frac{1}{j\omega C_p}I_1 + \left(\frac{1}{j\omega C_p} + \frac{1}{j\omega C_C} + R'_L\right)I_2 \tag{2}$$

From (1) and (2), I_2 is

$$I_2 = \frac{j\omega C_C}{\{1 - \omega^2(LC_C + LC_p + C_C C_p R_S R_L)\} + j\omega(C_C R_L - \omega^2 L C_C C_p R_L + C_C R_S + C_p R_S)} V_0 \tag{3}$$

By substituting this term into (1), I_1 is

$$I_1 = \frac{\left(\frac{C_C+C_p}{C_C} + j\omega C_p R_L\right)}{\left(\frac{C_C+C_p}{C_C}R_S - \omega^2 L C_p R_L + R_L\right) + j\left\{\omega C_p R_S R_L + \omega L \frac{C_C+C_p}{C_C} - \frac{1}{\omega C_C}\right\}} V_0 \tag{4}$$

Under the resonant condition, the corresponding impedance will be purely resistive. Thus, the imaginary part of the total impedance of the circuit equals zero and the resonant inductance L of the circuit can be calculated as

$$L = \frac{C_p R_L^2 + \frac{C_C+C_p}{\omega^2 C_C^2}}{\omega^2 C_p^2 R_L^2 + \left(\frac{C_C+C_p}{C_C}\right)^2} \tag{5}$$

From Figure 3, the stress voltage V_1 across the shield plate is found as

$$V_1 = V_0 - (R_S + j\omega L)I_1 \tag{6}$$

where V_0 is the input voltage, R_S is the ESR, $j\omega L$ is the inductor reactance, and I_1 is the current. Using (4), V_1 becomes

$$V_1 = \frac{R_L - j\frac{1}{\omega C_C}}{\left(\frac{C_C+C_p}{C_C}R_S - \omega^2 L C_p R_L + R_L\right) + j\left\{\omega C_p R_S R_L + \omega L \frac{C_C+C_p}{C_C} - \frac{1}{\omega C_C}\right\}} V_0 \tag{7}$$

2.2. Optimization for Minimal Loss by Impedance Matching

The complete system topology of Figure 3 is rearranged with R_0 as the internal characteristic resistance of the power source in Figure 5. An AC-input voltage is applied to the resonant-circuit input. In this analysis, we assume that the total impedance of the coupling interface and the secondary side of the shielded-CPT are matched to R_L'. To attain a maximum power transfer, the total reflected output impedance must equal the input impedance of the circuit [32]. This relationship is shown in (8):

$$R_0 = j\omega L + \frac{R_L'}{j\omega C R_L' + 1} \qquad (8)$$

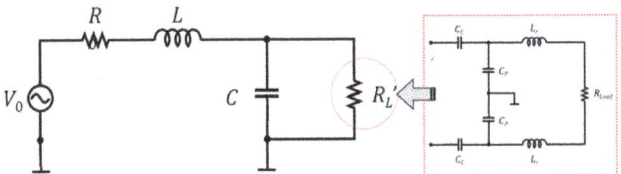

Figure 5. Circuit model for impedance analysis.

By multiplying both the numerator and the denominator by $(1 - j\omega C R_L)$, we have

$$R_0 = \frac{R_L'}{1 + \omega^2 C^2 R_L'^2} + j\omega \left(L - \frac{C R_L'^2}{1 + \omega^2 C^2 R_L'^2} \right) \qquad (9)$$

The reactive part of the circuit will be cancelled when it is in the resonant condition. Thus, the matching impedance state can be acquired as

$$R_0 = \frac{R_L'}{1 + \omega^2 C^2 R_L'^2} \qquad (10)$$

While the values of capacitance C and inductance L can be acquired from the quality factor Q of the components

$$C = \frac{1}{\omega R_L'} \sqrt{\frac{R_L'}{R_0} - 1} = \frac{Q}{\omega R_L'} \qquad (11)$$

$$L = \frac{C R_L'^2}{1 + \omega^2 C^2 R_L'^2} = \frac{R_0 Q}{\omega} \qquad (12)$$

The maximum value can be obtained using a derivative method. Therefore, the response to the load resistance R_L from (11) can be calculated by

$$\frac{dC}{dR_L'} = \frac{1}{\omega} \left(\frac{1}{R_L'^2} \sqrt{\frac{R_L'}{R_0} - 1} - \frac{1}{R_L'} \frac{1}{2R_0 \sqrt{\frac{R_L'}{R_0} - 1}} \right) \qquad (13)$$

$$\frac{dC}{dR_L'} = \frac{1}{\omega} \frac{1}{R_L'^2} \sqrt{\frac{R_L'}{R_0} - 1} \left(1 - \frac{R_L'}{2R_L' - 2R_0} \right) \qquad (14)$$

$$\frac{dC}{dR_L'} = 0 \qquad (15)$$

From (14) and (15), the relationship between the load resistance, R_L', and the internal characteristic resistance, R_0, can then be obtained as

$$R_L' = 2R_0 \qquad (16)$$

Thus, the optimum impedance-matching condition can be obtained when the load resistance, $R_L{'}$, is twice the internal characteristic resistance, R_0. Using the KCL method, the gain of the circuit in Figure 5 can be defined as

$$\frac{V_L}{V_0} = \frac{\frac{R_L'}{j\omega C R_L' + 1}}{R_0 + j\omega L + \frac{R_L'}{j\omega C R_L' + 1}} \tag{17}$$

Substituting (11) and (16) into (17), we obtain

$$\frac{V_L}{V_0} = \frac{1}{2} \frac{R_L'}{R_0 + jR_0 Q} \tag{18}$$

$$\left|\frac{V_L}{V_0}\right| = \frac{1}{2}\sqrt{\frac{R_L'^2}{R_0^2 + R_0^2 Q^2}} = \frac{1}{2}\sqrt{\frac{R_L'}{R_0}} \tag{19}$$

For analysis, the parameter of the circuit model is then defined to obtain a correlation in the capacitance, inductance, and circuit gain to the various load values. A square plate 10 cm in height and width achieves a coupling capacitance of 8.9 pF within 1 cm of the gap (see Table 1). The behaviors of the capacitance, inductance, and the circuit gain under various values of impedance are illustrated in Figures 6 and 7.

Table 1. Circuit parameters for ratio-of-resistance analysis.

Parameter	Value	Unit
e_0	8.85×10^{-14}	F/cm
l_C	10	cm
w_C	10	cm
d_C	1	cm
C	8.9×10^{-12}	F

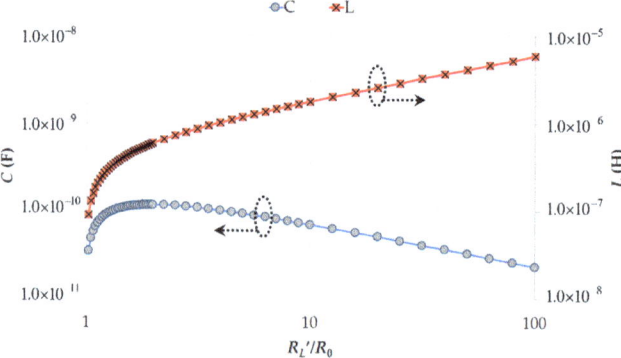

Figure 6. Capacitance and inductance behaviors under various impedance values.

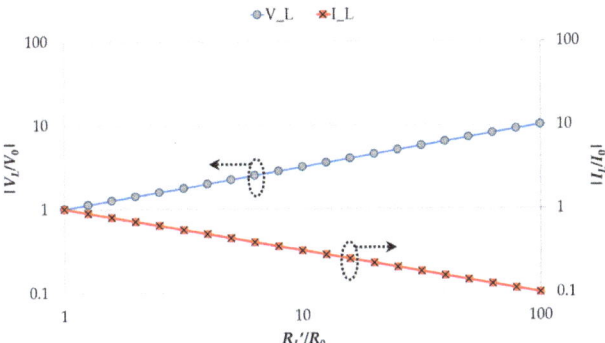

Figure 7. Voltage and current behaviors under various impedance values.

In Figure 6, the shield parasitic capacitance C_P becomes large to obtain a thin coupler unit. However, to develop a large ratio between the load resistance and the internal series resistance, R_L'/R_0, the shield capacitance C_P should be small. Thus, there is a trade-off relationship between the unit thickness and output voltage (Figure 7). With the increase of the load resistance, the load voltage, V_L, increases by a gradient of 1. By contrast, the load current decreases by gradient of −1. On the other hand, a large capacitance requires a small R_L' and the shield capacitance value is limited to 2 for R_L'/R_0. A double-LC-resonant-matching system (*LCLC* circuit) may solve this problem.

Figure 8 shows the relationship between the resistance ratio and power loss of the inductor. With increasing R_L', R_S rises because the inductor value increases, but I_L decreases. Thus, the loss in the inductor, P_S, decreases with increasing R_L' because P_S is given by $I_L^2 R_S$ and the gradients in the log-log plot are roughly 1/2 and −1 for R_S and I_L, respectively. Therefore, P_S decreases with increasing R_L' with a gradient of −3/2 in log-log plot (see Figure 9).

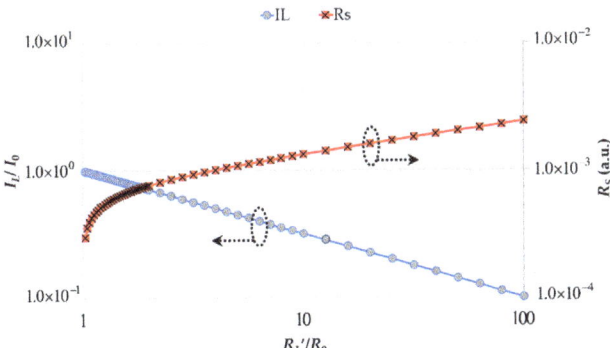

Figure 8. Current and equivalent series resistance (ESR) responses under various impedance values.

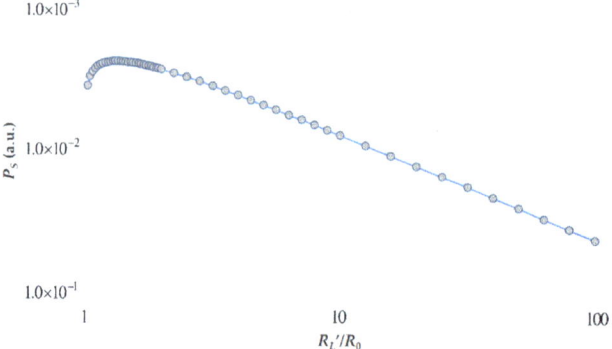

Figure 9. Power loss of the inductor under various impedance values.

2.3. Resonant Inductor Optimization for Minimal Loss

In this subsection, the ESR value of the inductor is investigated. Inductors can be constructed using many types of core materials; one is an air-core-type inductor that uses any non-magnetic material as its core to reduce core losses, i.e., eddy current & stray losses, especially when the operating frequency is very high. However, the use of a non-magnetic core also reduces inductance. Another type is a toroidal-core inductor, the core of which is made from a ferromagnetic material. The advantage of this circular core is that the magnetic field contains extremely low magnetic-flux leaks inside the core. The magnetic field at the core is higher because of a low leakage flow; hence, a toroidal-core inductor will have a higher inductance than a rod or bar-shaped core of the same material [32,33]. Figure 10 shows the dimensions and parameters of a toroidal-core inductor.

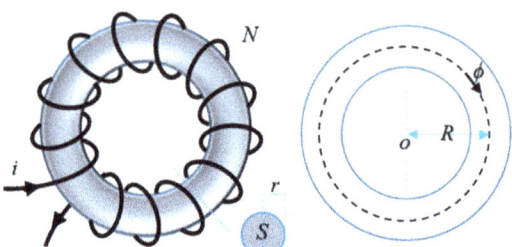

Figure 10. Inductor dimension.

The core inductance can be acquired as

$$L = \frac{N^2 \mu S}{l} = \frac{N^2 \mu (\pi r^2)}{2\pi R} = \frac{N^2 \mu r^2}{2R} \qquad (20)$$

where N is the number of turns, μ is the core-material permeability, S is the core thickness, l is the wire length, R and r represent the core radius and core thickness radius, respectively. The series resistance in the inductor is

$$R_S = \frac{\rho l_w}{S_w} = \frac{\rho(2\pi r N)}{\pi(\phi/2)^2} = \frac{8\rho r N}{\phi^2} \qquad (21)$$

$$L_S = \frac{N^2 \mu r^2}{2R}\left(\frac{\phi^4}{64\rho^2 r^2 N^2}R_S^2\right) = \frac{\mu \phi^4}{128\rho^2 R}R_S^2 \qquad (22)$$

Thus, the series resistance R_S can be written as

$$R_S = \frac{8\rho}{\phi^2}\sqrt{\frac{2RL_S}{\mu}} \qquad (23)$$

where ϕ and ρ represent the thickness and resistivity of the wire, respectively. The series inductance is

$$L_S = \frac{1}{\omega C_{2C}} \qquad (24)$$

Because the current flows through an inductor, its ESR consumes some power. Figure 11 shows the circuit model used to analyze the power loss in a resonant inductor without considering parasitic capacitance, CP. An AC voltage source is connected in series with the ESR RS, the inductance L_S, the coupling capacitance C_C, and the load resistance R_L.

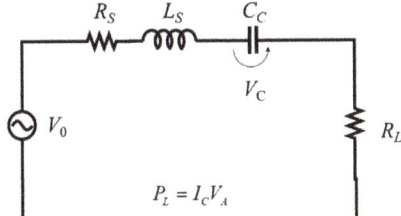

Figure 11. Circuit model for ESR-effect analysis.

The power loss of the impedance-matching inductor, P_S, is defined by

$$P_S = \frac{R_S}{R_L + R_S}P_{in} \qquad (25)$$

by substituting R_S from (23) and (24) into (25), P_S can be calculated as

$$P_S = \frac{1}{\frac{\omega R_L \phi^2}{8\sqrt{2}\rho}\sqrt{\frac{C_C \mu}{R}} + 1}P_{in} \qquad (26)$$

From (26), the relationship between the coupling capacitance, C_C, and the power loss of the matching inductor, P_S, is illustrated in Figure 12. Amplification of the coupling capacitance results in inductance drops. In addition, the ESR reduces and corresponds to the end product of (26), which is that the power loss decreases. However, because the distance between the couplers is fixed, the capacitance value is limited to a small number. Furthermore, the parasitic capacitance, C_P, will introduce a large value of capacitance that can affect the reduction of ESR, R_S.

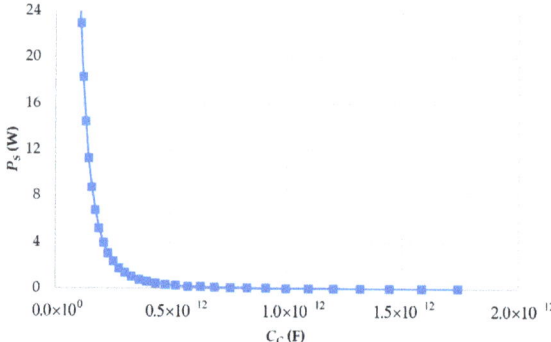

Figure 12. Inductance power loss as a function of coupling capacitance.

The effect of the parasitic capacitance, C_P, can be analyzed using (5). The increase of C_P occurs as the inductance value declines. From (23), the series inductor L_S proportionally influences the ESR, R_S, value. Its behavior is illustrated in Figure 13.

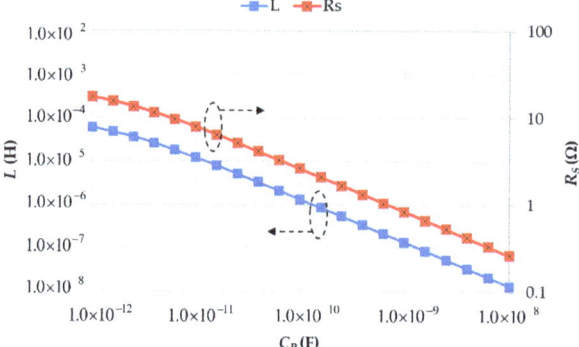

Figure 13. Series resistance of the resonant inductor with the change of the parasitic capacitance.

The relationship between power and efficiency with a frequency response is shown in Figure 14. It can be seen that, by having a lower ESR value (here below 1 Ω, see Figure 14a), less power is lost in the inductor, resulting in more power being delivered. The system may achieve an efficiency of over 95%. On the contrary, when the series resistance of the inductor is greater, it consumes some power, impacting the output power delivered to the load. From Figure 14b, the overall efficiency can be understood to drop. Thus, to minimize the loss caused by the resonant inductor, lower values of ESR result in higher efficiencies. One way to reduce ESR is by implementing a litz wire.

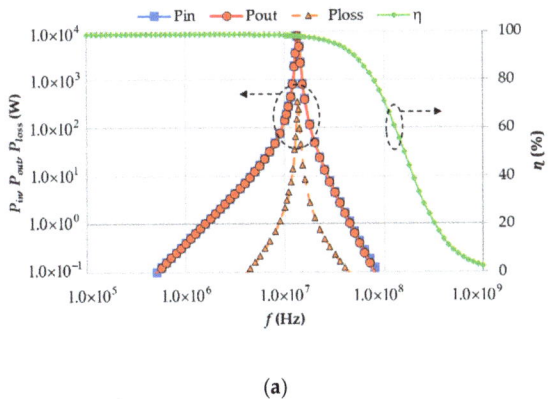

(a)

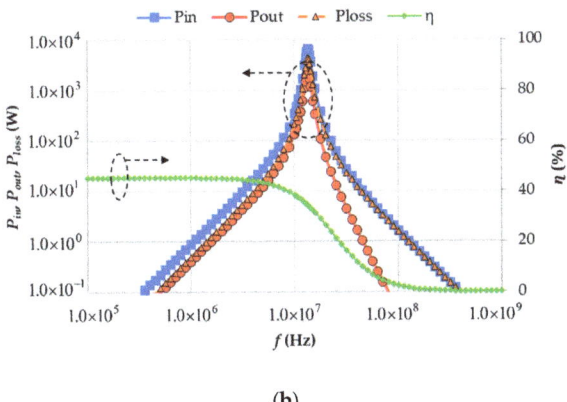

(b)

Figure 14. Frequency responses of power and efficiency: (**a**) Lower ESR value; (**b**) Higher ESR value.

2.4. Scaling Design for Various Loads

This subsection analyzes the behavior of power and efficiency under the variation of the load resistance. The resonant-inductance parameters used in this analysis are presented in Table 2. The ESR of the inductor, R_S, was calculated using (23). The value of inductance, L, changes due to parasitic capacitance, C_P. Furthermore, R_S depends on its value. The parameters of the circuit model for power and efficiency analysis are listed in Table 3.

Table 2. Parameters of the inductance model for power and efficiency analysis.

Parameter	Value	Unit
ρ	1.0×10^{-2}	W·cm
Φ	0.15	cm
R_{core}	2.5	cm
μ_o	1.26×10^{-8}	H/cm
μ_r	10×10^3	

Table 3. Parameters of circuit model for power and efficiency analysis.

Parameter	Value	Unit
C_C	1.33×10^{-12}	F
R_S	(23)	Ω
R_L	1 to 7.8×10^4	Ω
V_0	100	V
L	(5)	H
f	1.36×10^7	Hz
ω	8.5×10^7	rd/s

The behavior of the load resistance to the output power for various values of parasitic capacitance is illustrated in Figure 15a, while Figure 15b draws the load-resistance behavior to the efficiency with the shield parasitic-capacitance variations. The escalation in the load resistance R_L from 1 Ω to 125 Ω (which amplifies C_P from 1 pF to 1 nF) results in power increases. Increasing load resistance is determined to decrease power loss in the inductor.

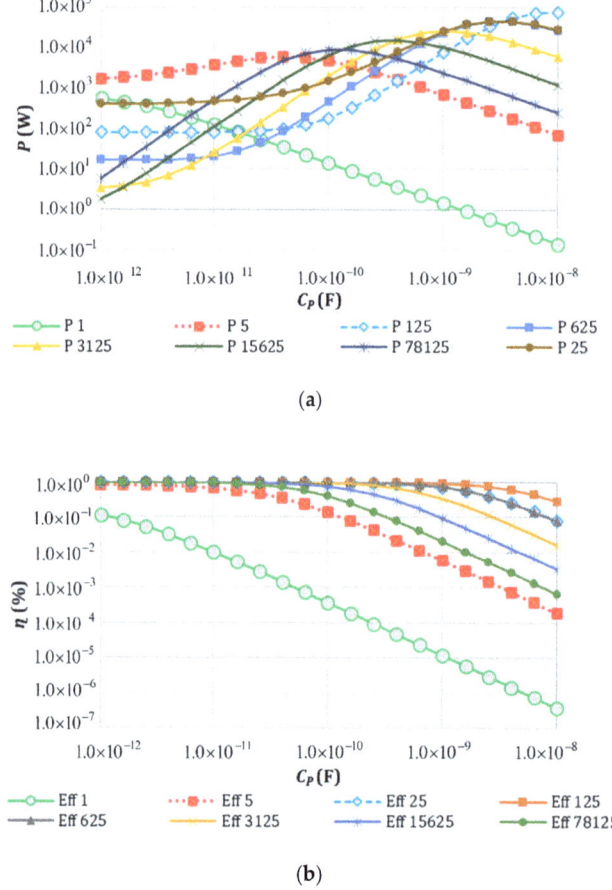

Figure 15. Effects of parasitic capacitance and load-resistance variations upon: (**a**) power behavior; (**b**) efficiency behavior.

2.5. Scaling Design for Various Parasitic Capacitances on the Shield-Coupler Stress Voltage

To acquire a thinner module of the shielded-CPT, the space gap between the shield plate and the coupler plate should be as short as possible. By decreasing the gap, the parasitic capacitance C_P increases. Assuming no stray capacitance on the edges of the plate, it can be estimated as

$$C_P = \frac{\varepsilon_r \varepsilon_0 S_P}{d_{SC}} \tag{27}$$

where d_{SC} and S_P are the gap distance between the shield-coupler plates and shield plate area, respectively. From (7), let us assume that C_C is much smaller than C_P; the stress voltage behavior is inversely proportional to the amount of parasitic capacitance (see (27)). While the gap is smaller, the parasitic capacitance will become greater. Here, increasing C_P will increase the stress voltage V_1. The EF value between the shield-coupler plates, E_{SC}, can be calculated as follows:

$$E_{SC} = \frac{V_1}{d_{SC}} \tag{28}$$

From (28), the EF level is amplified in proportion to the increase in stress voltage. As we know, the EF in air has a breakdown voltage E_{MAX} above 30 kV/cm [34], which is essential for system scaling. From this point, the EF strength between the shield-coupler plates should have a value under E_{MAX}. The effect of parasitic capacitance upon the stress voltage is shown in Figure 16. With decrease in d_{SC}, the capacitance C_P increases. The stress voltage over shield-coupler plates increased, which increased the EF between them proportionally.

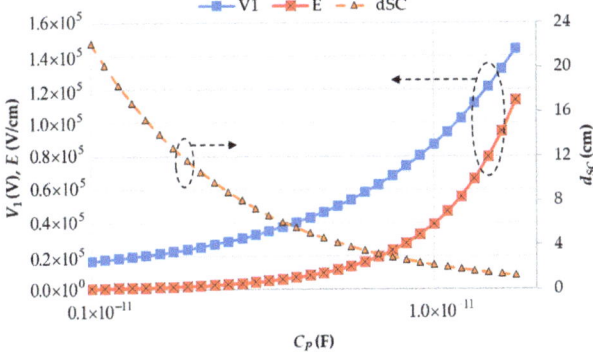

Figure 16. Stress voltage and EF level behavior over parasitic capacitance and gap variations.

3. Design Guidelines and Optimization of the Shielded-CPT System

The following example will demonstrate the design process of shielded-CPT to find optimum scaling and design values. The design step will rely upon the analytical approaches in the previous section. The scaling-factor of the shielded-CPT system will be described under the optimal condition. Several factors must be considered before the hardware is developed. A design guideline for scaling the system such that it meets constraints is shown in Figure 17. The individual unidentified-element values for this design example are L, R_S, and C_P. From (23), R_S is influenced by the L value. How L itself will vary depends on the parasitic capacitance C_P. At this point, we investigate C_P variations as describe in Section 2.5. With variation of C_P, L can be obtained using (5).

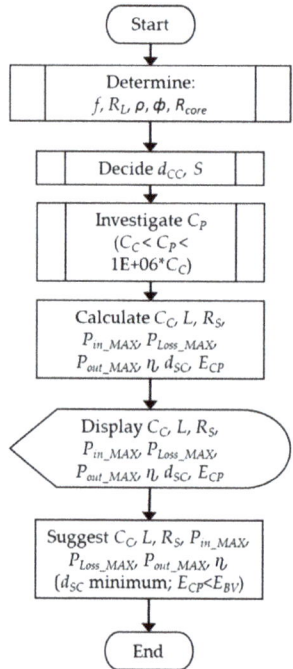

Figure 17. Design guidelines for optimizing the shielded-CPT system.

In this study, the shielded-CPT coupling interface need to transfer the power through the capacitive link at a frequency of 13.56 MHz. The frequency selection is decided by three factors. Firstly, the WPT systems for the high-power applications, requires not only to provide a high power through short-range of distance, but also a compact coupling interface that small enough to be integrated in the EV. The volume and weight of the resonant components are inversely proportional to the frequency. Thus, increasing the frequency by 10 s of MHz leads to weight lessening and power density enhancement [35]. Secondly, for the CPT system, the value of inductance is inversely proportional to the angular frequency in which the high frequency gives a low value of inductance, based on Equation (24). Thus, the inductor will become smaller for higher frequencies, offering an advantage to obtain a compact, lightweight, and small size CPT system. Thirdly, the limitation of the industrial, scientific, and medical (ISM) band for a MHz WPT. A fixed 6.78 MHz frequency as the lowest ISM band frequency is preferred. The international telecommunication union radio (ITU-R) communications sector currently recommends this single frequency on WPT for consumer devices because it has little or no negative impact on other licensed bands. A higher operating frequency in the ISM band, such as 13.56 MHz or 27.12 MHz, could further improve local freedom.

The shielded-CPT is connected to a 50 Ω resistive load. The capacitances of the coupling interface are separated by a distance of over 18 cm, which is the EV body-to-ground clearance requirement, with a plate are of 250 cm^2 corresponding to a main-coupling capacitance of approximately 1.11 pF. Let us assume that $\rho = 1.0 \times 10^{-4}$, where $\rho > \rho$ (Cu), $\phi = 0.15$ cm, and the radius of the air core $R = 2.5$ cm for the resonant inductor parameter. The designated parameters are listed in Table 4.

Table 4. Parameters of the shielded-CPT system for the design example.

Parameter	Value	Unit
ρ	1.0×10^{-4}	W·cm
ϕ	0.15	Cm
R_{core}	2.5	Cm
μ_o	1.26×10^{-8}	H/cm
μ_r	1.0×10^3	

The exemplary design results provide several options for the shielded-CPT parameter. Figure 18 shows L and R_S as functions of C_P. The inductance decreases in value (i.e., decreases its size) as the parasitic capacitance increases. Furthermore, the resistance value decreases. To obtain a small and compact module of the shielded-CPT, increasing the capacitance to a higher value is required. Furthermore, this increases power transfer through the couplers since a larger EF occurs.

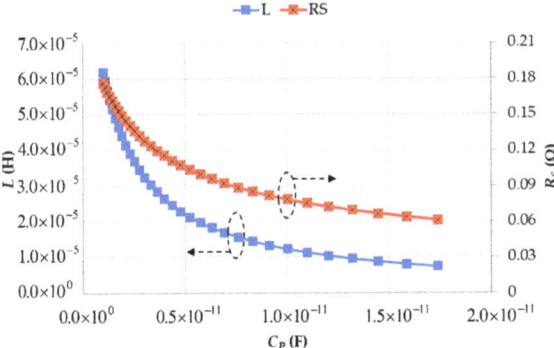

Figure 18. Inductance and series resistance characteristic for various parasitic-capacitance values.

On the contrary, more power transfer does not necessarily mean greater efficiency. Figure 19 illustrates the relationship among the parasitic capacitance to the power input, the power loss in the inductor, and the power output of the shielded-CPT. The efficiency decreases along with the gap between the plates. This occurs due to the increase in the current flowing through the primary-side components, meaning more input power is needed by the system. Thus, its efficiency is inversely proportionate to the compactness of the shielded-CPT module.

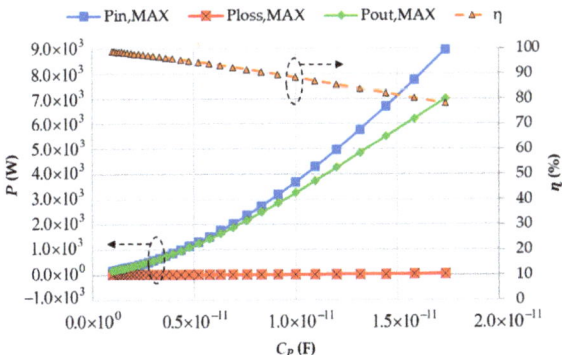

Figure 19. Power and efficiency characteristics for various parasitic-capacitance values.

The EF injected into the shield-coupler plates was limited to the breakdown voltage of air (30 kV/cm), as seen in Figure 20. In this example, at the limit point of 30 kV/cm, the parasitic-capacitance value for safety considerations must be below 2.78 pF. This condition can be obtained when the gap between the shield-coupler plates is much greater than 2.66 cm.

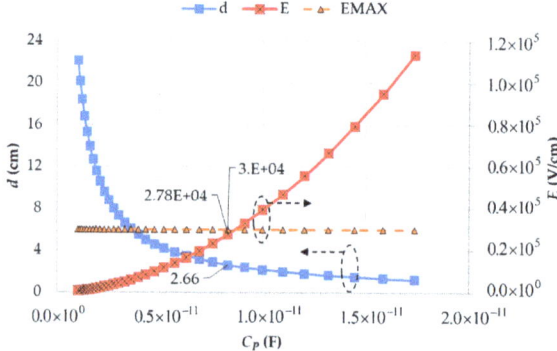

Figure 20. Gap and EF levels for various parasitic-capacitance values.

The minimum gap allowed in this design is shown in Figure 20. With a resonant frequency of 13.56 MHz and a coupler distance of 18 cm, the shielded-CPT coupling structure can be acquired for a gap of 2.92 cm, providing 7.59 pF of parasitic capacitance. 15.6 µH is needed as the total inductance works for resonance-circuit operation. For these parameters, the system is predicted to have an efficiency over 92%. The parameters of the shielded-CPT system in this example are listed in Table 5.

Table 5. Parameters of the shielded-CPT system for the design example.

Parameter	Value	Unit
d_{CC}	18	cm
d_{SC}	2.92	cm
C_C	1.11×10^{-12}	F
C_P	7.59×10^{-12}	F
R_S	0.09	Ω
R_L	50	Ω
V_0	50	V
L	1.56×10^{-5}	H
f	1.36×10^7	Hz

4. Hardware Implementation

The parameters obtained in the previous section are used in the hardware implementation. A single-layer PCB FR4 was used to create the coupling interface; the series-resonant inductors were fabricated manually and measured using an impedance network analyzer. The coupling interface of the shielded-CPT-system implementation is shown in Figure 21. The structure of the coupling capacitance follows the designated and optimized configuration.

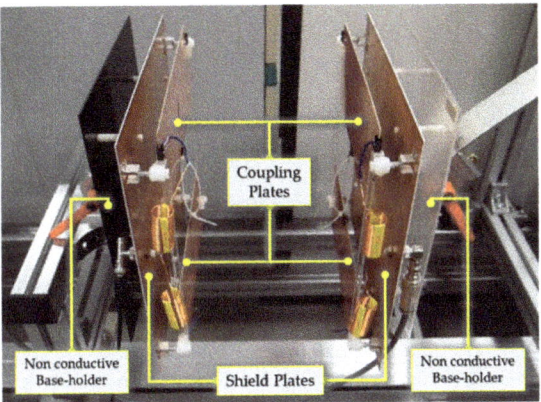

Figure 21. Hardware implementation of the shielded-CPT system.

Table 6 presents the hardware parameters of the shielded-CPT system for the design example. These were acquired from actual measurement of the system. In this implementation, the gap between the shield-coupler plates is about 3 cm. The resonant inductors are divided into four sections: upper and lower sides of the primary and secondary coupling interfaces. Some parameters exhibit differences from their calculated values due to manual production.

Table 6. Hardware parameters of the shielded-CPT system for the design example.

Parameter	Value	Unit
S_P	30.5 × 30.5	cm^2
S_C	10 × 25	cm^2
d_{CC}	18	cm
d_{CP}	3	cm
d_{CX}	10	cm
L_{t1}	7.86 × 10^{-6}	H
L_{t2}	7.65 × 10^{-6}	H
L_{r1}	8.2 × 10^{-6}	H
L_{r2}	7.6 × 10^{-6}	H

The efficiency of the coupling capacitance of the shielded-CPT system was measured using vector network analyzer. The coupling was connected to the measurement unit via a 50 Ω coaxial cable on both the transmitter and receiver sides. The measured S-parameter of the coupling is shown in Figure 22. The curve S [2,1] presents the power delivered from the transmitter to the receiver. Meanwhile, the curve S [1,1] presents the reflected power through the transmitter. At 13.56 MHz, the shielded-CPT in the hardware implementation has a delivered power above 0.86, equal to an efficiency above 86%.

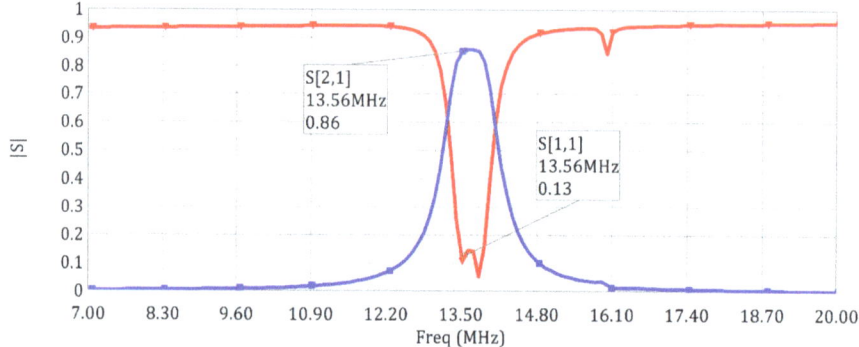

Figure 22. The gain of the hardware implementation of the shielded-CPT system.

A comparison between the design guidelines and the hardware implementation is presented in Table 7. The efficiencies of the design guidelines and the hardware implementation agree to within 93%. Since the implementation has a different parameter value, it affects the precision of the designated system.

Table 7. Comparison of results between the design guidelines and the hardware implementation.

Parameter	Value by Design	Value by Hardware	Unit	Similarity
S_P	10 × 25	30.5 × 30.5	cm^2	0.27
S_C	10 × 25	10 × 25	cm^2	1
d_{CC}	18	18	cm	1
d_{CP}	2.92	3	cm	0.98
d_{CX}	n.a.	10	cm	-
L_{t1+2}	15.6	15.5 × 10^{-6}	H	0.99
L_{r1+2}	15.6	15.8 × 10^{-6}	H	0.99
Eff	92	86	%	0.93

Figure 23 illustrates the EF emission that is analyzed using QuickField™ software. Most of the radiated EF appears between the coupling and shield plates over 20 kV/m. This condition deals with the calculated value of EF in Section 3 (Figure 20). However, the EF emission behind the shield plate is between 0 and 2 kV/m.

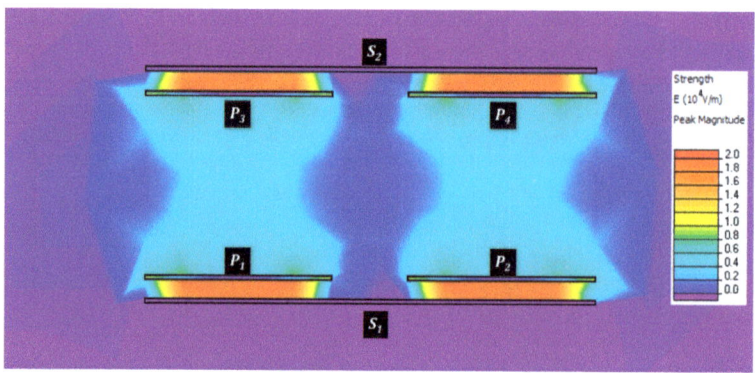

Figure 23. The EF emission of shielded-CPT system.

The highest EF denoted by red shows the value of 20 kV/m. Meanwhile, the lowest emission is shown by dark blue with an EF value between 0 and 2 kV/m. The ICNIRP guidelines and IEEE standards [36,37] mention 48.4 V/m and 60.75 V/m as the safety level of EF radiation for general public exposure limit under the operating frequency of 13.56 MHz. Figure 24 shows the EF emissions of the shielded-CPT system with a distance increase from (a) beside the coupling interface and (b) behind the shield plate. The emissions besides the coupling interface are measured over 1 kV/m. Its emissions are decreased to over 100 V/m when the distance is approximately 7.5 cm beside the coupler. Moreover, increasing the distance over 10 cm obtains a lower EF emission. However, the proposed shielded-CPT successfully reduced the emission behind the shield plate to below 10 V/m when the distance is less than 35 cm, which is reduced to below 1 V/m, meeting the ICNIRP and IEEE regulations.

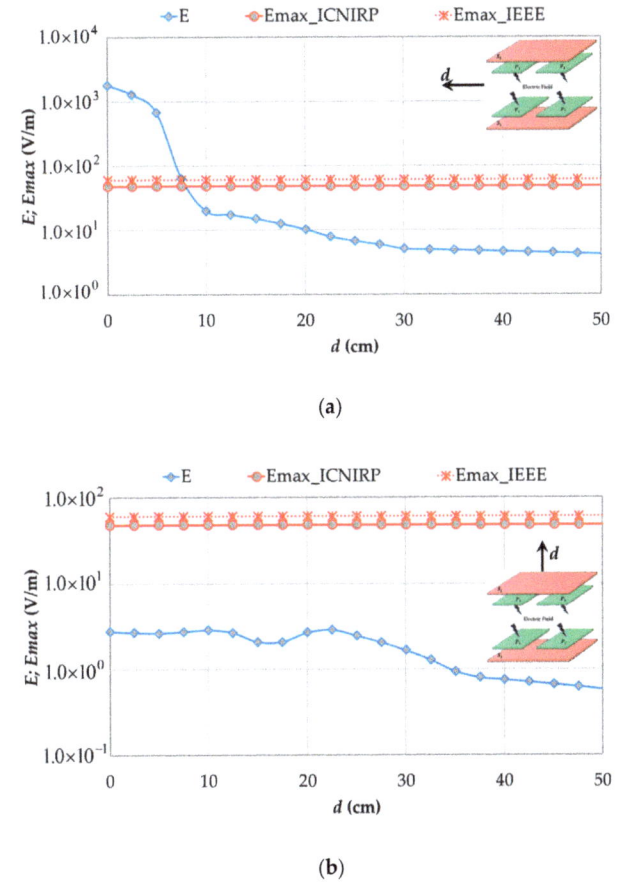

Figure 24. The EF emission of shielded-CPT system: (**a**) beside the coupling interface, (**b**) behind the shield plate.

5. Conclusions

This paper introduced scaling-factor and design guidelines for shielded-CPT. The theoretical design and analytical approach were described in detail with a simplified circuit model for analytical purposes.

From the EF and stress voltage of the shield-coupler plates, it was found that the stress voltage was proportional to the parasitic capacitance, C_P. The distance between the shield plate and coupler,

d_{SC}, needs to be minimized to obtain a thin and compact module. However, a larger C_P results in a strengthened stress voltage, which is limited to 30 kV/cm for safety considerations.

The power loss in the series inductor was investigated and found to be proportional to the inductance value. However, increasing C_P will reduce the value of L and result in low-power loss. The scaling used in the design was presented together with various load results; it was found that increasing R_L will decrease the inductor's power loss.

Process and procedure of scaling and designing the shielded-CPT was followed by analysis of the behavior of each factor, including voltage, current, power, parasitic capacitance, and system efficiency. Finally, the design guidelines for the shielded-CPT system were introduced. The design example for the hardware implementation was conducted successfully for the proposed method. It was found that, using these guidelines, an impressive hardware-parameter calculation and implementation was obtained. Thus, the proposed method can be recommended for designing shielded-CPT systems with scaling-factor and safety consideration.

Author Contributions: Conceptualization, A.M. and R.H.; methodology, A.M. and R.H.; validation, A.M. and S.A.; visualization, A.M.; writing—original draft preparation, A.M.; writing—review and editing, S.A. and R.H.; supervision, R.H. All authors have read and agreed to the published version of the manuscript.

Funding: This research received no external funding.

Acknowledgments: The authors would like to thank Mitsuru Masuda and the members of Automotive System and Device Laboratories, Furukawa Electric Co., Ltd. through their technical support. The Ministry of Research and Technology, Republic of Indonesia through the RISET-Pro scholarship support to A.M. (World Bank Loan No. 8245-ID).

Conflicts of Interest: The authors declare no conflict of interest.

References

1. Dai, J.; Ludois, D.C. Capacitive Power Transfer through a Conformal Bumper for Electric Vehicle Charging. *IEEE J. Emerg. Sel. Top. Power Electron.* **2016**, *4*, 1015–1025. [CrossRef]
2. Huang, L.; Hu, A.P.; Swain, A.; Dai, X. Comparison of two high frequency converters for capacitive power transfer. In Proceedings of the 2014 IEEE Energy Conversion Congress and Exposition (ECCE), Pittsburgh, PA, USA, 14–18 September 2014; pp. 5437–5443.
3. Yusop, Y.; Saat, S.; Ghani, Z.; Husin, H.; Nguang, S.K. Capacitive power transfer with impedance matching network. In Proceedings of the 2016 IEEE 12th International Colloquium on Signal Processing & Its Applications (CSPA), Malacca, Malaysia, 4–6 March 2016; pp. 124–129.
4. Funato, H.; Kobayashi, H.; Kitabayashi, T. Analysis of transfer power of capacitive power transfer system. In Proceedings of the 2013 IEEE 10th International Conference on Power Electronics and Drive Systems (PEDS), Kitakyushu, Japan, 22–25 April 2013; pp. 1015–1020.
5. Lu, F.; Zhang, H.; Hofmann, H.; Mi, C. A Double-Sided LCLC Compensated Capacitive Power Transfer System for Electric Vehicle Charging. *IEEE Trans. Power Electron.* **2015**, *30*, 6011–6014. [CrossRef]
6. Li, C.; Zhao, X.; Liao, C.; Wang, L. A graphical analysis on compensation designs of large-gap CPT systems for EV charging applications. *CES Trans. Electr. Mach. Syst.* **2018**, *2*, 232–242. [CrossRef]
7. Lu, J.; Zhu, G.; Lin, D.; Zhang, Y.; Jiang, J.; Mi, C.C. Unified Load-Independent ZPA Analysis and Design in CC and CV Modes of Higher Order Resonant Circuits for WPT Systems. *IEEE Trans. Transp. Electrif.* **2019**, *5*, 977–987. [CrossRef]
8. Vu, V.B.; Dahidah, M.; Pickert, V.; Phan, V.T. An Improved LCL-L Compensation Topology for Capacitive Power Transfer in Electric Vehicle Charging. *IEEE Access* **2020**, *8*, 27757–27768. [CrossRef]
9. Regensburger, B.; Kumar, A.; Sinha, S.; Xu, J.; Afridi, K.K. High-Efficiency High-Power-Transfer-Density Capacitive Wireless Power Transfer System for Electric Vehicle Charging Utilizing Semi-Toroidal Interleaved-Foil Coupled Inductors. In Proceedings of the 2019 IEEE Applied Power Electronics Conference and Exposition (APEC), Anaheim, CA, USA, 17–21 March 2019; pp. 1533–1538.
10. Estrada, J.; Sinha, S.; Regensburger, B.; Afridi, K.; Popovic, Z. Capacitive wireless powering for electric vehicles with near-field phased arrays. In Proceedings of the 2017 47th European Microwave Conference (EuMC), Nuremberg, Germany, 10–12 October 2017; pp. 196–199.

11. Mostafa, T.M.; Muharam, A.; Hattori, R. Wireless battery charging system for drones via capacitive power transfer. In Proceedings of the 2017 IEEE PELS Workshop on Emerging Technologies: Wireless Power Transfer (WoW), Chongqing, China, 20–22 May 2017; Volume 3, pp. 1–6.
12. Park, C.; Park, J.; Shin, Y.; Huh, S.; Kim, J.; Ahn, S. Separated Circular Capacitive Couplers for Rotational Misalignment of Drones. In Proceedings of the 2019 IEEE Wireless Power Transfer Conference (WPTC), London, UK, 18–21 June 2019; pp. 635–638.
13. Vincent, D.; Huynh, P.S.; Patnaik, L.; Williamson, S.S. Prospects of Capacitive Wireless Power Transfer (C-WPT) for Unmanned Aerial Vehicles. In Proceedings of the 2018 IEEE PELS Workshop on Emerging Technologies: Wireless Power Transfer (Wow), Montréal, QC, Canada, 3–7 June 2018; pp. 1–5.
14. Zou, L.J.; Hu, A.P.; Su, Y. A single-wire capacitive power transfer system with large coupling alignment tolerance. In Proceedings of the 2017 IEEE PELS Workshop on Emerging Technologies: Wireless Power Transfer (WoW), Chongqing, China, 20–22 May 2017; pp. 93–98.
15. Theodoridis, M.P. Effective capacitive power transfer. *IEEE Trans. Power Electron.* **2012**, *27*, 4906–4913. [CrossRef]
16. Mi, C. High power capacitive power transfer for electric vehicle charging applications. In Proceedings of the 2015 6th International Conference on Power Electronics Systems and Applications (PESA), Hong Kong, China, 15–17 December 2015; pp. 1–4.
17. Cao, C.; Wang, L.; Chen, B.; Harper, J.; Bohn, T.; Dobrzynski, D.; Hardy, K. Real-Time Modeling to Enable Hardware-in-the-Loop Simulation of Plug-In Electric Vehicle-Grid Interaction. In Proceedings of the 13th ASME/IEEE International Conference on Mechatronic and Embedded Systems and Applications; American Society of Mechanical Engineers, Cleveland, OH, USA, 6–9 August 2017; Volume 9, pp. 1–10.
18. Das, H.S.; Rahman, M.M.; Li, S.; Tan, C.W. Electric vehicles standards, charging infrastructure, and impact on grid integration: A technological review. *Renew. Sustain. Energy Rev.* **2020**, *120*, 109618. [CrossRef]
19. Wang, L.; Cao, C.; Chen, B. Grid-Tied single-phase Bi-directional PEV charging/discharging control. *SAE Int. J. Passeng. Cars-Electron. Electr. Syst.* **2016**, *9*, 275–285. [CrossRef]
20. Faria, R.; Moura, P.; Delgado, J.; de Almeida, A.T. Managing the Charging of Electrical Vehicles: Impacts on the Electrical Grid and on the Environment. *IEEE Intell. Transp. Syst. Mag.* **2014**, *6*, 54–65. [CrossRef]
21. Muharam, A.; Mostafa, T.M.; Nugroho, A.; Hapid, A.; Hattori, R. A Single-Wire Method of Coupling Interface in Capacitive Power Transfer for Electric Vehicle Wireless Charging System. In Proceedings of the 2018 International Conference on Sustainable Energy Engineering and Application (ICSEEA), Tangerang, Indonesia, 1–2 November 2018; pp. 39–43.
22. Lu, F.; Zhang, H.; Mi, C. A Two-Plate Capacitive Wireless Power Transfer System for Electric Vehicle Charging Applications. *IEEE Trans. Power Electron.* **2018**, *33*, 964–969. [CrossRef]
23. Zhang, H.; Lu, F.; Hofmann, H.; Liu, W.; Mi, C.C. Six-plate capacitive coupler to reduce electric field emission in large air-gap capacitive power transfer. *IEEE Trans. Power Electron.* **2018**, *33*, 665–675. [CrossRef]
24. Muharam, A.; Masuda, M.; Hattori, R.; Hapid, A. Compactly Assembled Transmitting and Receiving Modules with Shield for Capacitive Coupling Power Transfer System. In Proceedings of the 2019 IEEE PELS Workshop on Emerging Technologies: Wireless Power Transfer (WoW), London, UK, 18–21 June 2019; pp. 257–262.
25. Masuda, M. A high electric power supply to electric cars using the electric field resonance. *Furukawa Rev.* **2018**, *49*, 23–31.
26. Zhang, H.; Lu, F.; Hofmann, H.; Liu, W.; Mi, C.C. A 4-Plate Compact Capacitive Coupler Design and LCL-Compensated Topology for Capacitive Power Transfer in Electric Vehicle Charging Applications. *IEEE Trans. Power Electron.* **2016**, *31*, 1. [CrossRef]
27. Lee, J.-B.; Baek, J.-I.; Kim, J.-K. A New Zero-Voltage Switching Half-Bridge Converter With Reduced Primary Conduction and Snubber Losses in Wide-Input-Voltage Applications. *IEEE Trans. Power Electron.* **2018**, *33*, 10419–10427. [CrossRef]
28. Arteaga, J.M.; Aldhaher, S.; Kkelis, G.; Yates, D.C.; Mitcheson, P.D. Design of a 13.56 MHz IPT system optimised for dynamic wireless charging environments. In Proceedings of the 2016 IEEE 2nd Annual Southern Power Electronics Conference (SPEC), Auckland, New Zealand, 5–8 December 2016; pp. 1–6.
29. Narayanamoorthi, R.; Vimala Juliet, A.; Bharatiraja, C.; Padmanaban, S.; Leonowicz, Z.M. Class E power amplifier design and optimization for the capacitive coupled wireless power transfer system in biomedical implants. *Energies* **2017**, *10*, 1409. [CrossRef]

30. Choi, U.-G.; Yang, J.-R. A 120 W Class-E Power Module with an Adaptive Power Combiner for a 6.78 MHz Wireless Power Transfer System. *Energies* **2018**, *11*, 2083. [CrossRef]
31. Muharam, A.; Ahmad, S.; Hattori, R.; Obara, D.; Masuda, M.; Ismail, K.; Hapid, A. An Improved Ground Stability in Shielded Capacitive Wireless Power Transfer. In Proceedings of the 2019 International Conference on Sustainable Energy Engineering and Application (ICSEEA), Tangerang, Indonesia, 23–24 October 2019; pp. 1–5.
32. Bowick, C.; Blyler, J.; Ajluni, C. *RF Circuit Design*; John Wiley & Sons, Inc.: Hoboken, NJ, USA, 2008; ISBN 9780470405758.
33. Behagi, A. Resonant Circuits and Filters. In *RF And Microwave Circuit Design: A Design Approach Using (ADS)*; Techno Search: Ladera Ranch, CA, USA, 2015; pp. 1–68. ISBN 0890069735.
34. Masuda, M.; Kusunoki, M.; Obara, D.; Nakayama, Y.; Hamada, H.; Negami, S.; Kaizuka, K. Wireless power transfer via electric coupling. *Furukawa Rev.* **2013**, *44*, 33–38.
35. Choi, J.; Tsukiyama, D.; Tsuruda, Y.; Davila, J.M.R. High-Frequency, High-Power Resonant Inverter With eGaN FET for Wireless Power Transfer. *IEEE Trans. Power Electron.* **2018**, *33*, 1890–1896. [CrossRef]
36. International Commission on Non-Ionizing Radiation Protection (ICNIRP). Guidelines for Limiting Exposure to Electromagnetic Fields (100 kHz to 300 GHz). *Health Phys.* **2020**, *118*, 483–524. [CrossRef] [PubMed]
37. IEEE Standards Coordinating Committee 39. *IEEE Standard for Safety Levels with Respect to Human Exposure to Electric, Magnetic, and Electromagnetic Fields, 0 Hz to 300 GHz*; IEEE Std C95.1-2019 (Revision IEEE Std C95.1-2005/ Inc. IEEE Std C95.1-2019/Cor 1-2019); IEEE: Piscataway, NJ, USA, 2019; pp. 1–312.

© 2020 by the authors. Licensee MDPI, Basel, Switzerland. This article is an open access article distributed under the terms and conditions of the Creative Commons Attribution (CC BY) license (http://creativecommons.org/licenses/by/4.0/).

Article

A Resonant Coupling Power Transfer System Using Two Driving Coils

Changping Li [1], Bo Wang [2], Ruining Huang [3] and Ying Yi [2],*

[1] College of Communication and Information Engineering, Chongqing University of Posts and Telecommunications, Chongqing 400065, China
[2] Division of Information and Computing Technology, College of Science and Engineering, Hamad Bin Khalifa University, Education City 34110, Qatar
[3] School of Mechanical Engineering and Automation, Harbin Institute of Technology, Shenzhen 518055, China
* Correspondence: yyi@hbku.edu.qa

Received: 17 June 2019; Accepted: 22 July 2019; Published: 29 July 2019

Abstract: This paper presents a resonance-based wireless power transfer (R-WPT) system using two multi-layer multi-turn inductor coils on the transmission side and a third coil on the receiver side. We theoretically characterized and optimized the system in terms of quality factor (Q factor) of the coils and power transfer efficiency (PTE). In our R-WPT prototype, the alternating currents (AC) were simultaneously applied to two transmitter coils, which, in turn, transferred power wirelessly to the secondary coil with a 3-mm radius on the receiving end. Owing to the optimization of the inductive coils, all of the coils achieved the highest Q-factor and PTE at the resonance frequency of 2.9 MHz, and the transfer distance could be extended up to 30 mm. The results show that the PTE was greater than 74% at a separation distance of 5 mm and about 38.7% at 20 mm. This is distinctly higher than that of its 2 and 3-coil counterparts using only one driving coil.

Keywords: resonance-based wireless power transfer (R-WPT); resonance frequency; power transfer efficiency (PTE); 3-coil system

1. Introduction

Wireless power transfer (WPT) circuits have been widely deployed in applications such as implantable electronics [1–4] and biomedical treatment systems [5,6]. It enables a miniaturized system design, as well as a battery-less operation. The WPT systems also exhibits great potential to be combined with wireless communication electronics for data transmission [7–9]. In particular, resonance-based WPT (R-WPT) systems [10–12] that use resonant coupling coils can achieve high (e.g., 78.6% at distance of 10 mm [11]) power transfer efficiency (PTE) [13,14]. Different from an inductive coupling power transfer technique [15], the R-WPT utilizes a capacitor and an inductor to form a LC resonant circuit. The driving and the load coils operate at the same resonant frequency and form a resonant coupling, which allows maximum power to be delivered wirelessly from the driver to the load [13]. The PTE of the R-WPT can be typically improved by optimizing the quality factor Q and the structure of the driving coils [2].

From a different perspective, the R-WPT systems do face some challenges, especially when they are used to power implantable electronics. In that case, the size of the implanted coil should be minimized, which makes it difficult to obtain a maximum PTE at the target resonance frequency [12]. Meanwhile, due to an increased transfer distance, a reduction of PTE can also be expected. Designing all coils to achieve their peak Q factors at the resonance frequency, is a feasible approach to improve the PTE when the geometry of the implantable coil is restricted [2,10]. In addition to the Q factor, the optimization of the coil structures can also improve the PTE. For example, a 4-coil resonant system has a higher PTE than its 2-coil counterpart at a relatively large coil separation distance [12]. However, the device becomes bulkier, and a driving coil with a high Q-factor is required.

Recently, the 3-coil R-WPT system is attracting growing interest due to its high PTE and the maximum amount of power delivered to the load (PDL) [11]. One structure of the 3-coil R-WPT systems uses one primary coil at the transmitter side and one secondary (or intermediate [16]) and one load coil at the receiver side ("1T-2R" for simplicity). This system can a high PTE while it is too bulky as the three coils are separated from each other [11,16,17]. Another type of the 3-coil R-WPT system uses one driving, one primary and one load coil (2T-1R) [17,18], which can minimize the load coil size but exhibits a low PTE (e.g., 17% at distance of 15 mm [10]). In this paper, we propose a new R-WPT structure by using two driving and one load coils. In the structure, the two driving coils are wound together to effectively enhance the coupling coefficient between the driving and the load coils. Moreover, all coils are designed to achieve their peak Q factors at the resonance frequency to further increase the PTE of the proposed system. As a result, the proposed design can achieve an improved PTE, as well as meeting the strict size requirements of implantable electronics.

This paper is organized as follows: Section 2 describes the theoretical basis of R-WPT and presents the proposed 3-coil model in terms of inductance, capacitance, and the Q-factor; Section 3 explains the PTE based on a circuit-based schematic diagram and evaluates simulation results and experimental measurement; Section 4 provides conclusions.

2. Theory and Design

2.1. Theoretical Basis

It is known that an alternating current (AC) applied to an inductor coil can induce a varying magnetic field, which can induce an AC on its neighboring inductor coil. Different from previous WPT systems, in which the AC is applied to only one coil at the transmitter side, in this work, as shown in Figure 1, two driving coils simultaneously carried AC to enhance the coil-induced magnetic field. A COMSOL simulation model was used to verify the concept of the proposed two driving coil design, the radius of the driving and implant coils were 0.6 and 0.3 cm, respectively. These parameters followed the dimensions of the coil prototypes in the experiments. In the simulation model, a relatively small AC of 0.025 mA was applied to two driving coils. The distribution of the magnetic field on the coil surface is presented in Figure 1. The small coil, a wireless power receiver, was placed 10 mm away from the driving coils. As presented in the simulation data, the electric field distribution on the receiver coil confirmed that the power was wirelessly delivered.

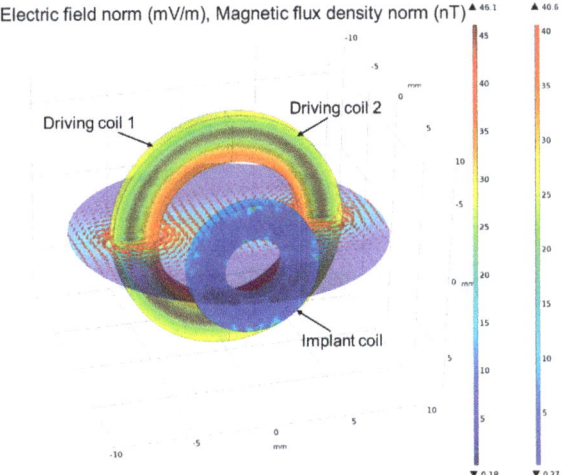

Figure 1. Illustration of an electric field induced by a varying magnetic field.

2.2. Proposed Coil Model

The driving and load coils of the proposed R-WPT system were fabricated using an AWG46 Litz wire. The long electrically conductive wire was wound into a structure with N_a layers and N_t turns (Figure 2 (left)). The Litz wire can mitigate the negative impacts on the skin, and the proximity effects [10], because it consists of multiple thin wire strands that are twisted together and electrically insulated from each other (Figure 2 (right)). In this work, the working frequency (*f*) of 2.9 MHz was selected (to be discussed in Figure 4 and Section 3), such that the skin depth could be calculated as 38 µm according to $\sqrt{2/2\pi f \mu_0 \sigma}$, where σ is the conductivity of the wire and μ_0 is the permeability of free space. The diameter of the wire strand was 39.8 µm, which revealed the suitability of the wire to the working frequency. The other physical parameters of the wire were listed in Table 1 and used in the system simulations.

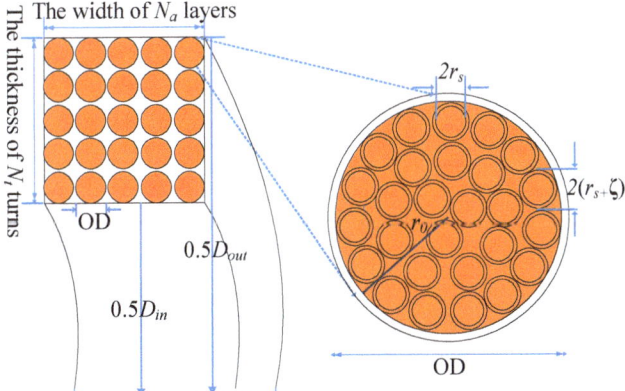

Figure 2. The driving coils' cross section (D_{in}: Inner conductor loop, D_{out}: Outer conductor loop, OD: Outer diameter) with multi-layer and multi-turn structure (**left**); cross sections of single turn with multi-strand wires (**right**).

Table 1. Litz wire property.

Radius of wire per strand, r_s	19.9 [μm]
Number of strands, N_s	20
Area efficiency, β	55%
Conductivity, σ	58 [S/mm^2]
Isolation Thickness, ζ	3 [μm]
Inner radius, r_0	110 [μm]
Relative permittivity, ε_r	3

According to the structure shown in Figure 2, the coil's total self-inductance is [10,11]:

$$L_{self} = \sum_{l=1}^{N_a} \left\{ \sum_{i=1}^{N_t} L(a_i, R) + \sum_{i=1}^{N_t} \sum_{\substack{j=1 \\ j \neq i}}^{N_t} M(a_i, a_j, \rho = 0, d = d_{ij}) \right\} \quad (1)$$

The first and second terms represent the summation of each turn's inductance and the summation of mutual inductance between each turn, respectively. They can be expressed as follows [10]:

$$\begin{cases} L(a, R) = a\mu_0 [In(\frac{8a}{R}) - 2], \quad (R \ll a) \\ M(a, b, \rho = 0, d) = \mu_0 \sqrt{ab} [(\frac{2}{k} - k)K(k) - \frac{2}{k}E(k)] \\ k = [\frac{4ab}{(a+b)^2 + d^2}]^{1/2} \end{cases} \quad (2)$$

where a_i is the radius of the ith turn of a coil, R is the wire radius, N_t is the total turns on each coil layer, d_{ij} is the relative distance between ith turn and jth turn. $\rho = 0$ means that the turns on the layer are perfectly aligned. $K(k)$ and $E(k)$ are the complete elliptic integrals of the first and second kind, respectively. In the simulation model, for the driving and the receiver coils, N_t was 52 and 55, respectively. The simulated inductances of coils were calculated as shown in Figure 3. It can be observed that the coil's inductance slightly increases with frequency for the ranges of 1 MHz to 5 MHz, in which the working frequency was located (to be explained in Figure 4).

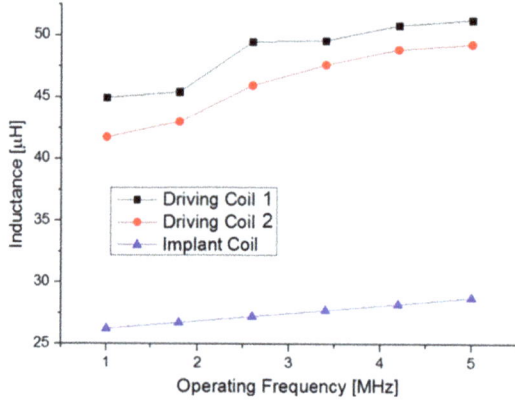

Figure 3. Inductance simulation of driving coils and implant coil with multi-layer and multi-turn versus different frequency.

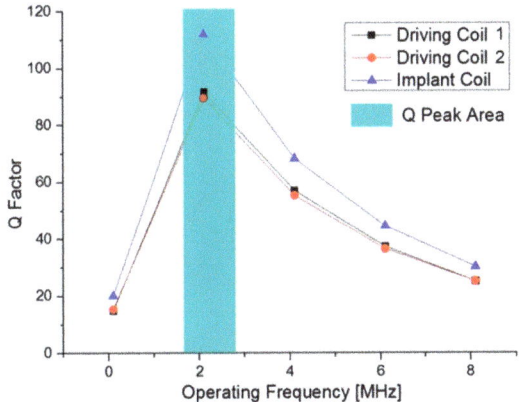

Figure 4. Q-factors' curves versus operating frequency.

In our work, the wire was tightly wound into multi-turn multi-layer, so the parasitic capacitance should be taken into account. In our coil design, the parasitic capacitance can be simplified to:

$$C_{par} = \frac{1}{N_t^2}[C(N_t - 1)] \quad (3)$$

where C is the parasitic capacitance between neighboring turns and it is given by [11]:

$$C = \varepsilon_0 \varepsilon_r \int_0^{\pi/4} \frac{\pi D r_0}{\varsigma + \varepsilon_r r_0 (1 - \cos\theta) + 0.5\varepsilon_r d} d\theta \quad (4)$$

where D, r_0, ς, d, ε_r and ε_0 are the average coil diameter, inner radius of each bunch, thickness of the insulation layer (Figure 2 (right)), relative distance between neighboring turns, relative permittivity of the insulation and the dielectric constant of the free space, respectively.

The Q-factor of the inductor is critical for the WPT [2,10]. Achieving a peak Q-factor in the WPT system can bring an improved PTE, it also reduces heat dissipation, which is especially important for implantable medical devices (IMDs), as a significant temperature change can induce organ or tissue burning. The Q-factor of an inductor is defined as:

$$Q = 2\pi f L / R \quad (5)$$

where f is the operating frequency, L is the inductance of the coil, and R is its effective AC resistance. Figure 4 shows the simulated Q-factor of the three coils versus the operating frequency. The highlighted area represents the frequency range where the coils achieved the maximum Q-factor, this is denoted as f_{peak}. As shown in Figure 4, all coils obtained their peak Q-factor within the frequency range of 2 to 3 MHz.

Table 2 lists the simulation results of the inductance and Q-factor given by Figures 3 and 4. A precision impedance analyzer (Agilent 4294A) was used to validate simulation results of the coils, with their measurement also given in Table 2. Based on the measured coils' inductances, three capacitors of 110 pF, 110 pF and 200 pF for driving coils #1, #2, and implantable coil #3, respectively, were employed to form the three LC resonators. Finally, all coils were tuned to operate at the same resonance frequency. Based on Equation (6) and circuit model, as shown in Figure 5a, a resonant frequency of 2.2 MHz can be expected for the designed R-WPT system. This frequency was located in the f_{peak} (2–3 MHz as depicted by Figure 4), and it can also be regarded as a safe electromagnetic frequency exposed to the human body according to the international commission on non-Ionizing radiation protection (ICNIRP) and the IEEE standards [19,20].

$$f_o = \frac{1}{2\pi \sqrt{L_m C_m}} \quad (6)$$

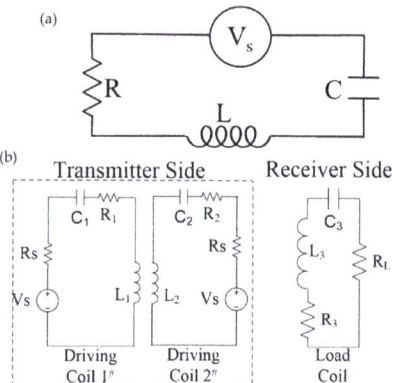

Figure 5. (a) Basic coil lumped circuit; (b) circuit model of the 3-coil R-WPT system.

Table 2. Experimental verification of coils' Q factor and inductance: (M) denotes measurement; and (S) denotes simulation result.

Coil Num.	Inductance M (2.5 MHz)	Inductance S (2.5 MHz)	Resistance M (2.5 MHz)	Q-factor M (2.5 MHz)	Q-factor S (2.5MHz)
1	55 uH	49 uH	11 Ω	82	90
2	53 uH	46 uH	11 Ω	78	87
3	34 uH	27 uH	5 Ω	95	112

3. Results and Discussions

3.1. Power Transfer Efficiency

For the conventional WPT systems, the two-coil system using the primary (Q_p) and the secondary coils (Q_s) is commonly seen. The Q-factor of the coils and the coupling coefficient (k) determine the PTE, which is given by [12]:

$$\eta_{2-coil} = \frac{k^2 Q_p Q_s}{1 + k^2 Q_p Q_s} \quad (7)$$

The coil lumped circuit is modeled as shown in Figure 5a, the total impedance around the resonant circuit is:

$$Z = R + j\omega L + 1/j\omega C \quad (8)$$

where R, L, and C are the inductance, resistance, and capacitance of the circuit, and j is the imaginary constant. According to Kirchoff's law, the induced current in the resonator is:

$$I = M \frac{dI_e}{dt} / Z \quad (9)$$

where I_e is the current applied in the driving coil, t is time. M is the mutual inductance between the driver and the load coils, which is expressed as:

$$M = k\sqrt{L_d L_l} \quad (10)$$

Compared to the one driving coil systems, the direction of the AC passing through the two driving coils must be kept the same. Otherwise, the effective magnetic field may be mutually canceled and weakened. The schematic diagram of the 3-coil R-WPT system is shown in Figure 5b, where V_s is the power source applied to the primary coil and Rs is the source impedance. In this system, the induced current in the load coil L_3 can be expressed as:

$$I_3 = \left(M_{13} \cdot \frac{dI_1}{dt} + M_{23} \cdot \frac{dI_2}{dt}\right)/|Z_3| \qquad (11)$$

From Equations (9) and (11), the induced current in the proposed two driving coil R-WPT system is much larger than that of the conventional WPT system given the same input power.

3.2. Simulation Results

Adopting the multi-turn multi-layer structure as shown in Figure 2, we wound 52 turns for both two driving coils and 55 turns for the receiver coil, respectively. Although increasing the number of the turns may increase the Q-factor of the coil, note that the AC resistance also gets increased accordingly, bringing negative influence on the Q-factor. In our design, a peak Q-factor was observed experimentally when the number of turns was around 54, while the Q-factor dropped when the number of turns was over 54. The detailed dimensions and geometric specifications of the three coils are given in Table 3. The driving and the load coils (functioned as the implantable coil) were concentric. The induced current was verified via COMSOL simulation using the coils' physical parameters listed in Tables 1 and 3. In this simulation, the distance between the driving and implant coils was 5 mm. The applied AC was 0.1 mA with a frequency of 2.2 MHz to each driving coils, Figure 6 depicts the cross section of the induced current density at the receiver side. This simulation result clarifies that power can be transferred wirelessly through the resonant system. Moreover, we can observe that the inner turns of the coil received more power than that of the outer of the coil, which is a reasonable outcome according to magnetic field theory. The ratio of the induced current to the applied current was about 12% with a transfer distance of 5 mm.

In the next simulation, the influence of distance and orientation to the PTE were investigated. In the simulation model, two co-axial driving coils and the load coil were positioned at a distance of 5 mm apart, the center axis of the driving coils were aligned with that of the load coil, we calculated the PTE, then the transfer distance was increased up to 25 mm, the correlated PTE versus distance were given in Figure 7a. A maximum PTE of 80% was achieved at a separation of 5 mm. As a comparison, same simulation conditions were applied to the 2T-1R model using only one driving coil. As expected, our design shows better PTE than its counterpart due to an increased mutual inductance between the driving and load coils. Locating the load coil 5 mm apart from the driving coils, we horizontally rotated the load coil with respect to the axis of driving coils, this may likely reflect the real scenario of wirelessly powering the implants. The PTE calculated versus the rotation angle were shown in Figure 7b. The shift and the rotation reduce the magnetic flux through the load coil, resulting in a decreasing coupling coefficient between the driving and the load coils. Consequently, the PTE drops with an increasing rotation angle.

Table 3. Coils' physical specification by measurements.

Type	Coil Num.	Outer Dia. (mm)	Inner Dia. (mm)	Turn/Layers N_t	Layers N_a	DC Resistance (Ω)	Capacitance (pF)
Driving Coil	1	21	12	13	4	2.2	110
Driving Coil	2	21	12	13	4	2.5	110
Load Coil	3	12	6	11	5	1.8	200

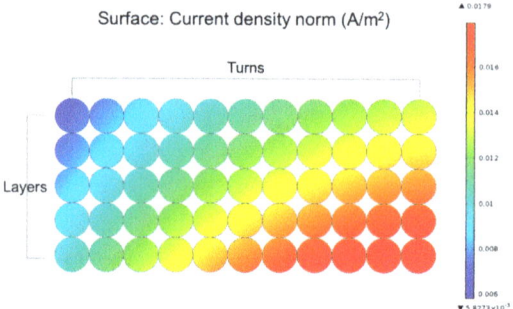

Figure 6. Cross section of surface current density of implant coil in COMSOL simulation when the transfer distance is 5 mm.

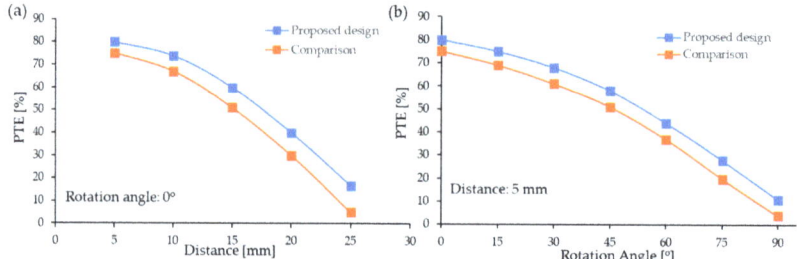

Figure 7. PTE calculated versus (**a**) transfer distance and (**b**) rotation angle in the simulation.

3.3. Experimental Measurements

To demonstrate the validity of the proposed two driving coil R-WPT technique, a measurement setup was implemented as shown in Figure 8 (left). An NI XI 5402 was used to apply 1 mW AC power to the driving coils, which were located on the opposite side of the load coil (the inset of Figure 8 (left)). With a transfer distance of d_{23} = 20 mm, the measured voltage waveforms of the driving and the implant coils were shown in Figure 8 (right). The received power was obtained by measuring the induced current and the voltage directly on the load coil. When the signal generator was tuned to f_0 = 2.94 MHz, the induced voltage on the load coil exhibited a peak value (V_{pp}) of 5.52 V. In Figure 8 (right), the voltage of the driving coil was much lower than that of the load coil. The reason could mainly be that the number of turns of the load coil was larger than that of the driving coil [12]. Though, the induced current across the load coil was I_{pp} = 0.56 mA, much lower than the applied current on the driving coils in our measurements. The corresponding PTE was 38% (derived by the multiplication of the induced current I_{rms} and voltage V_{rms} over the emitted power). Moreover, a phase difference between the input and output voltage waveforms was observed, this is because the coupling between the driving and the load coils may shift the original phase if the multiple transmitters are used [17]. As shown in Figure 9 (red dot), the PTE was measured at a different transfer distance d_{23} from 5 mm to 30 mm. The highest PTE is 74% at the distance of 5 mm (equal to near tissue thickness). Moreover, the PTE can keep a high value (38.7%) over a distance up to 20 mm (equal to deep tissue thickness). Table 4 summarizes the parameters of our proposed system, as well as the comparison with previously reported works in the literature.

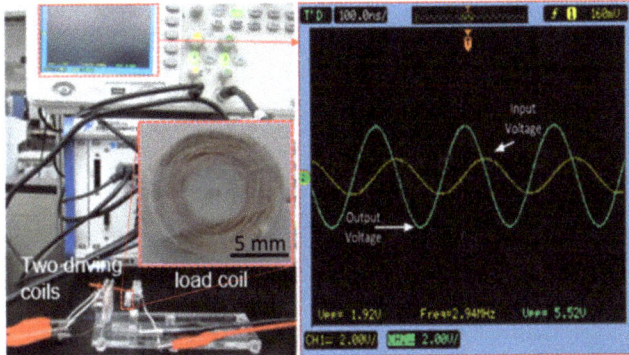

Figure 8. The experimental setup for the three-coil WPT system (**left**); The measurement results of the input voltage (on driving coil) and output voltage (on load coil), d_{23} = 20 mm (**right**).

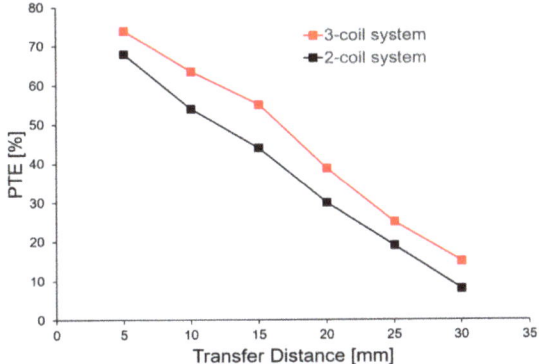

Figure 9. PTE measurements for the proposed 3-coil system and typical 2-coil system.

Table 4. Comparison with previous works.

Ref.	Design	Size (cm)	Frequency (MHz)	Distance ($\lambda \cdot 10^{-3}$)	PTE
[10]	2T-1R	Tx: $\pi \times 1.52$ Rx: $\pi \times 1.752$	6.76	0.34	17%
[11]	1T-2R	Tx: $\pi \times 2.152$ Rx: $\pi \times 0.52$	13.56	0.45	78.6%
[17]	1T-2R	Tx: 35×30 Rx: 31.5×22.5	0.66	0.35	59.7%
This work	2T-1R	Tx: $\pi \times 0.62$ Rx: $\pi \times 0.32$	2.9	0.05	74%
This work	2T-1R	Tx: $\pi \times 0.62$ Rx: $\pi \times 0.32$	2.9	0.2	38%

As a comparison, a typical 2 coil system with an equivalent number of turns of the driving coils was also tested. Kept the same experimental setup, the AC power was applied to the driving coil (104 turns) and measured the PTE of the same load coil versus the distance, the results are represented by the black dot in Figure 9. As expected, we can clearly observe that our 3-coil system, with two driving coils and one load coil shows much better PTE than the 2-coil counterpart. This is because all the coils achieved their peak Q-factors in our design, while the Q-factor of the driving coil may deviate from its peak value in the 2-coil model, resulting in a lower PTE. Similar observations and results can be also found in [11], however, our work provided a simple load coil structure instead of designing two separate coils on the receiver side [11], which makes the implantable applications bulky and complicated.

4. Conclusions

This work presents a 3-coil R-WPT using a pair of driving coils which is intended to provide an improved PTE. The geometries of the coils are optimized through simulations, consequently, the coils achieve their peak Q-factors at the resonant frequency. Moreover, the influence of the transfer distance and orientation on the PTE is demonstrated. The results indicate that the proposed design provides a higher PTE than its 3-coil counterpart using only one driving coil in both cases. Finally, the performance of the proposed design is further validated via experiments, and the results show that the 3-coil system using two driving coils provides higher PTE compared to the 2-coil system with an equivalent geometry setting. Besides the PTE, the resonant frequency (or the operating frequency) is another critical factor for implantable applications. The selection of the frequency must be carefully considered in order to avoid safety concerns and electromagnetic wave interference. This will need in-depth and careful studies and further refinement efforts and will be part of our future work following the current feasibility study.

Author Contributions: Conceptualization, C.L. and Y.Y.; methodology, Y.Y.; validation, Y.Y. and R.H.; formal analysis, C.L. and B.W.; writing—original draft preparation, C.L.; writing—review and editing, C.L., B.W., R.H. and Y.Y.

Funding: This publication was made possible by NPRP grant NPRP11S-0104-180192 from the Qatar National Research Fund (a member of Qatar Foundation). The statements made herein are solely the responsibility of the authors.

Conflicts of Interest: The authors declare no conflicts of interest.

References

1. Yi, Y.; Zaher, A.; Yassine, O.; Kosel, J.; Foulds, I. A remotely operated drug delivery system with an electrolytic pump and a thermo-responsive valve. *Biomicrofluidics* **2015**, *9*, 052608. [CrossRef] [PubMed]
2. Yang, Z.; Liu, W.; Basham, E. Inductor modeling in wireless links for implantable electronics. *IEEE Trans. Magn.* **2007**, *43*, 3851–3860. [CrossRef]
3. Yi, Y.; Buttner, U.; Foulds, I. A cyclically actuated electrolytic drug delivery device. *Lab Chip* **2015**, *15*, 3540–3548. [CrossRef] [PubMed]
4. Xue, R.F.; Cheng, K.W.; Je, M. High-Efficiency Wireless Power Transfer for Biomedical Implants by Optimal Resonant Load Transformation. *IEEE Trans. Circuits Syst. I Regul. Pap.* **2013**, *60*, 867–874. [CrossRef]
5. Yi, Y.; Chen, J.; Selvaraj, M.; Hsiang, Y.; Takahata, K. Wireless Hyperthermia Stent System for Restenosis Treatment and Testing with Swine Model. *IEEE Trans. Biomed. Eng.* **2019**, (in press).
6. Yi, Y.; Huang, R.; Li, C. Flexible substrate-based thermo-responsive valve applied in electromagnetically powered drug delivery system. *J. Mater. Sci.* **2019**, *54*, 3392–3402. [CrossRef]
7. Baker, M.; Sarpeshkar, R. Feedback Analysis and Design of RF Power Links for Low-Power Bionic Systems. *IEEE Trans. Biomed. Circuits Syst.* **2007**, *1*, 28–38. [CrossRef] [PubMed]
8. Han, L.; Li, L. Integrated wireless communications and wireless power transfer: An overview. *Phys. Commun.* **2017**, *25*, 555–563. [CrossRef]
9. Catrysse, M.; Hermans, B.; Puers, R. An inductive power system with integrated bi-directional data-transmission. *Sens. Actuators A Phys.* **2004**, *115*, 221–229. [CrossRef]
10. Yi, Y.; Buttner, U.; Fan, Y.; Foulds, I. Design and optimization of a 3-coil resonance-based wireless power transfer system for biomedical implants. *Int. J. Circuit Theory Appl.* **2014**, *43*, 1379–1390. [CrossRef]
11. Kiani, M.; Jow, U.M.; Ghovanloo, M. Design and Optimization of a 3-Coil Inductive Link for Efficient Wireless Power Transmission. *IEEE Trans. Biomed. Circuits Syst.* **2011**, *5*, 579–591. [CrossRef] [PubMed]

12. RamRakhyani, A.; Mirabbasi, S.; Chiao, M. Design and Optimization of Resonance-Based Efficient Wireless Power Delivery Systems for Biomedical Implants. *IEEE Trans. Biomed. Circuits Syst.* **2011**, *5*, 48–63. [CrossRef] [PubMed]
13. Kurs, A.; Karalis, A.; Moffatt, R.; Joannopoulos, J.; Fisher, P.; Soljacic, M. Wireless Power Transfer via Strongly Coupled Magnetic Resonances. *Science* **2007**, *317*, 83–86. [CrossRef] [PubMed]
14. Yi, Y.; Kosel, J. A remotely operated drug delivery system with dose control. *Sens. Actuators A Phys.* **2017**, *261*, 177–183. [CrossRef]
15. Jow, U.; Ghovanloo, M. Design and Optimization of Printed Spiral Coils for Efficient Transcutaneous Inductive Power Transmission. *IEEE Trans. Biomed. Circuits Syst.* **2007**, *1*, 193–202. [CrossRef] [PubMed]
16. Kim, J.W.; Son, H.C.; Kim, K.H.; Park, Y.J. Efficiency Analysis of Magnetic Resonance Wireless Power Transfer With Intermediate Resonant Coil. *IEEE Antennas Wirel. Propag. Lett.* **2011**, *10*, 389–392. [CrossRef]
17. Ahn, D.; Hong, S. Effect of Coupling Between Multiple Transmitters or Multiple Receivers on Wireless Power Transfer. *IEEE Trans. Ind. Electron.* **2013**, *60*, 2602–2613. [CrossRef]
18. Arakawa, T.; Coguri, S.; Krogmeier, J.V.; Kruger, A.; Love, D.J.; Mudumbai, R.; Swabey, M.A. Optimizing Wireless Power Transfer from Multiple Transmit Coils. *IEEE Access* **2018**, *6*, 23828–23838. [CrossRef]
19. Park, S.; Kim, M. Numerical Exposure Assessment Method for Low Frequency Range and Application to Wireless Power Transfer. *PLoS ONE* **2016**, *11*, e0166520. [CrossRef] [PubMed]
20. Christ, A.; Douglas, M.; Nadakuduti, J.; Kuster, N. Assessing Human Exposure to Electromagnetic Fields from Wireless Power Transmission Systems. *Proc. IEEE* **2013**, *101*, 1482–1493. [CrossRef]

© 2019 by the authors. Licensee MDPI, Basel, Switzerland. This article is an open access article distributed under the terms and conditions of the Creative Commons Attribution (CC BY) license (http://creativecommons.org/licenses/by/4.0/).

Article

Coupling Coefficient Measurement Method with Simple Procedures Using a Two-Port Network Analyzer for a Multi-Coil WPT System

Seon-Jae Jeon and Dong-Wook Seo *

Department of Radio Communication Engineering, Korea Maritime and Ocean University (KMOU), 727 Taejong-ro, Yeongdo-gu, Busan 49112, Korea; seonjae@kmou.ac.kr
* Correspondence: dwseo@kmou.ac.kr; Tel.: +82-51-410-4427

Received: 10 September 2019; Accepted: 16 October 2019; Published: 17 October 2019

Abstract: In this paper, we propose a measurement method with a simple procedure based on the definition of the impedance parameter using a two-port network analyzer. The main advantage of the proposed measurement method is that there is no limit on the number of measuring coils, and the method has a simple measurement procedure. To verify the proposed method, we measured the coupling coefficient among three coils with respect to the distance between the two farthest coils at 6.78 and 13.56 MHz, which are frequencies most common for a wireless power transfer (WPT) system in high-frequency band. As a result, the proposed method showed good agreement with results of the conventional S-parameter measurement methods.

Keywords: coupling coefficient; impedance matrix; multiple coils; mutual inductance; scattering matrix; transfer impedance; wireless power transfer

1. Introduction

The conventional wireless power transfer (WPT) system with two coils has obvious limits; the system is very sensitive to the transmission distance and alignment between the coils [1,2]. To resolve these problems, adaptive or tunable matching networks have been adopted [3,4], and a transmitting module not with a single coil, but with multiple coils has been also introduced into the WPT system [5,6]. Especially, the magnetic beamforming, which focuses magnetic fields from the multiple transmitting coils to the receiving coil, has recently attracted lots of attention. For an ideal multi-coil WPT system, the coupling coefficients among transmitting coils should be zero, but in reality, they are not. The non-zero coupling coefficient among transmitting coils is one of the major causes of deterioration of the power transfer efficiency (PTE) in the magnetic beamforming. To minimize the effect of non-zero coupling coefficients, the phase and magnitude of signals input to transmitting coils should be adjusted to the optimum values that are estimated from the measured coupling coefficients among coils. The adjustment makes the all current of transmitting coils in phase, and results in a maximum receiver current and maximum PTE [5,6]. Therefore, the accurate and simple procedures for measurement of the coupling coefficient among multiple coils is essential for implementing the magnetic beamforming of a multi-coil WPT system.

For a WPT system with an operating frequency of a low-frequency (LF) band such as 110 to 205 kHz, the coupling coefficient (or mutual inductance) is usually measured by an LCR meter [7] or an impedance analyzer [8]. On the other hand, in a high-frequency (HF) band such as 6.78 or 13.56 MHz, most coils have a frequency-dependent characteristic due to the effect of ac resistance and parasitic elements. Therefore, a vector network analyzer is commonly used to measure the coupling coefficient in HF bands [9]. In the case of multi-coil WPT systems, the measurement issue arises primarily from the common vector network analyzer (VNA) having only two ports. Although a multiport VNA can measure the coupling coefficient among a greater number of coils, it is still only available for coils with equal or less than the number of ports.

In this paper, we simulate the effect of inaccurately measured coupling coefficients on the PTE of a multi-coil WPT system, and propose a method to readily measure the coupling coefficients among multiple coils using a two-port VNA based on the definition of the impedance matrix. In addition, this measurement method can overcome the limitation of the performance of a multi-port VNA for more coils than the number of ports. We verify the validity of the proposed method from the results that the proposed method has the same performance as conventional measurement methods according to distance.

2. Effect of Inaccurate Coupling Coefficient on PTE of a Multi-Coil WPT System

In this section, we simulate the PTE of a multi-coil WPT system with magnetic beamforming by applying inaccurate coupling coefficient information to examine the importance of an accurate coupling coefficient measurement.

Consider the multiple-input single-output (MISO) WPT system configured as four transmitting coils and a single receiving coil, where all transmitting coils are modeled as an inductance of 1.1 µH and a parasitic resistance of 0.4 ohm. Additionally, we assume that all coupling coefficients among transmitting coils are fixed to 0.05, and coupling coefficients between transmitting and receiving coils are also fixed to 0.05. The assumption that all coupling coefficients among transmitting coils are the same for all transmitting coils is difficult to realize in practice, but this assumption is effective in simulating the PTE of a multi-coil WPT system, because the input signals of all transmitting resonators have the same value when the magnetic beamforming algorithm is applied. Figure 1 shows the schematic diagram of the MISO WPT system for circuit simulation using the Keysight's ADS (Advanced Design System).

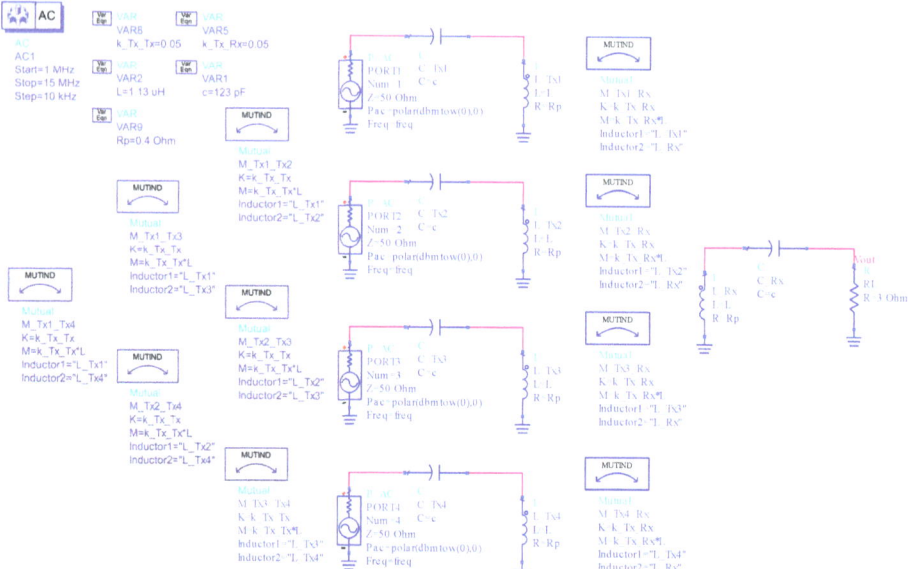

Figure 1. Schematic diagram of a multiple-input single-output (MISO) wireless power transfer (WPT) system for the advanced design system (ADS) circuit simulation.

To apply magnetic beamforming to the MISO WPT system, the voltage of transmitting coils must be given by [6]:

$$\vec{v}_T^{bf} = \left(\mathbf{Z}_T + \frac{\omega^2 \mathbf{M}^T \mathbf{M}}{R_R}\right) \vec{i}_T^{bf}, \qquad (1)$$

where $\mathbf{M}$ is the mutual inductance matrix between the transmitting and receiving coils; R_R is the total resistance of the receiving resonator; $\vec{i}_T^{bf}$ is the transmit current and is proportional to $\mathbf{M}$; and $\mathbf{Z}_T$ is the inter-coupling matrix among transmitting coils and is constructed as the coupling coefficients (or mutual inductances) among transmitting coils. That is, the amplitude and phase of the transmit voltage is dominantly determined by coupling coefficient information. Therefore, the more accurate the coupling coefficient is, the more accurately the transmit voltage for the magnetic beamforming is obtained.

By applying the coupling coefficient with error term to the magnetic beamforming, we simulated the PTE degradation of the MISO WPT system at 13.56 MHz. The simulated results are shown in Figure 2. When the exact coupling coefficient of 0.05 is used for the beamforming, the PTE achieves about 77%. As the error rate of the coupling coefficient increases, the PTE decreases. Coupling coefficient error rates of 10% and 20% result in PTE degradations of 7% and 15%, respectively, compared with no error. Since the coupling coefficient among coils usually has a low level with 10^{-2} order, error rate of 10% means 10^{-3} order that is quite challenging to measure. Additionally, the larger the number of coils used in the HF band, the more difficult and cumbersome it is to measure the coupling coefficients. Therefore, a highly accurate measurement method of coupling coefficients with simple procedures is very important to achieve the ideal PTE of a multi-coil WPT system.

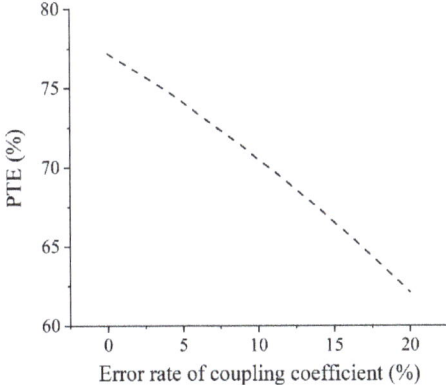

Figure 2. Simulated power transfer efficiency (PTE) with respect to error rate of coupling coefficients between coils in the MISO WPT system.

3. Conventional and Proposed Methods

This section presents conventional and proposed measurement methods of coupling coefficients among multiple coils. For the measurement of coupling coefficients, we use some assumptions. First, the measured coupling coefficient is not by electrical coupling but by magnetic coupling. In the case of the radio-frequency identification (RFID) with a few millimeters distance between coils, electrical coupling should be considered [10]. However, a WPT system in HF band with the transmission distance of several tens of centimeters usually considers only magnetic coupling. Second, the measured mutual inductance is not the ideal inductance, but the mutual inductance affected by the frequency and surrounding objects by parasitic elements. This is because the values used for the magnetic beamforming of the multi-coil WPT system are the mutual inductance affected by parasitic elements.

3.1. Conventional Measurement Methods

In HF bands, the conventional methods to measure coupling coefficients among three or more coils can be categorized into two types, namely, those that use multi-port and those that use two-port VNAs. The former method simply measures the $n \times n$ [S] matrix for n coils that are directly connected to ports of a multi-port VNA without an additional resonant capacitor. Then, the measured [S] matrix is converted into the [Z] matrix on an external PC. Consequently, the coupling coefficient between the ith and jth coils is obtained by:

$$k_{ij} = \frac{M_{ij}}{\sqrt{L_i \cdot L_j}} = \frac{\text{Im}\{Z_{ij}\}}{\sqrt{\text{Im}\{Z_{ii}\} \cdot \text{Im}\{Z_{jj}\}}}, \qquad (2)$$

where M_{ij} is the mutual inductance between the ith and jth coils, and L_i and L_j are the self-inductances of the coils, respectively. Here, Z_{ij} is the ij entry of the [Z] matrix, and it is often referred to as the transfer impedance for $i \times j$. The other conventional method repeatedly measures 2×2 [S] matrices for all possible sets of two coils, where the other coils must be terminated with a 50 ohm load from the definition of the S parameter, and then synthesize the $n \times n$ [S] matrix for the total n coils. This method is theoretically equivalent to the former method [11,12].

3.2. Proposed Measurement Method

From Equation (2), it is confirmed that the coupling coefficient is calculated using only elements of the [Z] matrix without the need for the [S] matrix. Furthermore, the calculation of k_{ij} requires only impedance parameters related to the ith and jth coils. The Z parameters, often called open-circuit impedance parameters, are measured or calculated by applying current to one port and measuring the resulting voltages at all other ports opened. That is, to obtain the impedance parameters related the ith and jth coils, the others port must be open-circuited. In addition, the required Z parameters can be obtained by converting the measured 2×2 [S] matrix into the 2×2 [Z] matrix.

Consider a transmitter array configured as four coils, as shown in Figure 3. To measure the coupling coefficient (k_{14}) between coils 1 and 4, both the coils 1 and 4 are connected to a two-port network without any resonant capacitor, and all other coils are open-terminated. Because most VNAs provide a function to mathematically manipulate measured data, such as Keysight's equation editor, we can immediately check the coupling coefficient between coils 1 and 4 on the display of the two-port VNA by converting a 2×2 [S] matrix to a 2×2 [Z] matrix and applying Equation (2).

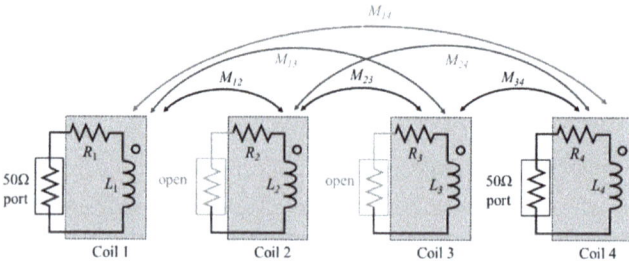

Figure 3. Circuit diagram for measuring coupling coefficient between coils 1 and 4 for four coils using a two-port vector network analyzer (VNA).

3.3. Simplicity of the Proposed Method

Figure 4 shows the measurement procedures of the proposed method and two conventional methods for coupling coefficients among multiple coils. Method I using a multi-port VNA is very convenient because all relationships among coils are measured at once. However, if the number of coils is larger than the number of ports of the network analyzer, this method cannot be applied. Additionally, the measured high-order [S] matrix is not instantaneously converted to the corresponding [Z] matrix on the common VNA, so the [S] matrix should be saved in an SnP (touchstone) file format and converted on an external PC. Method II using a two-port VNA requires repetitive two-port measurements $_nC_2$ times for all possible sets of two coils and all unconnected coils terminated with a 50 ohm load. In addition, the total $n \times n$ [S] matrix should be synthesized using 2×2 [S] matrices. On the other hand, in the proposed method, only two coils are connected with a two-port VNA, and the coupling coefficient obtained from Equation (2) can be immediately checked on the display of the VNA according to the distance between coils. Additionally, this method is not limited to the number of coils. Therefore, the proposed method has very simple procedures.

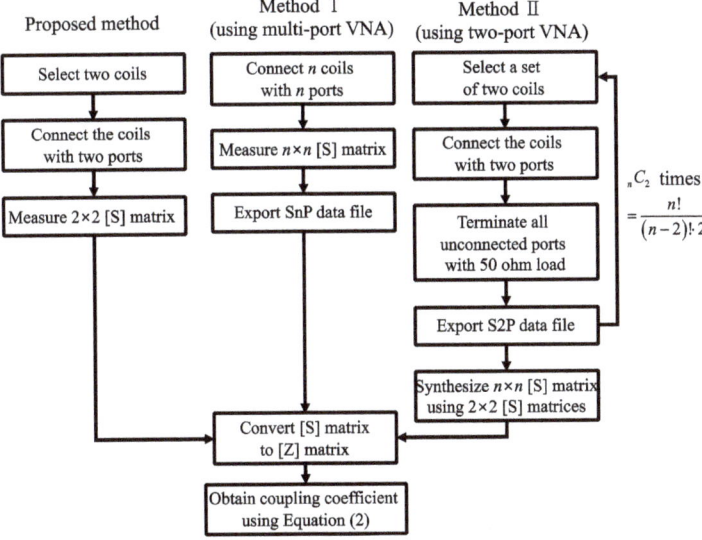

Figure 4. Measurement procedures for the proposed and conventional methods.

4. Experimental Setup and Results

This section presents experimental setup and results to validate the proposed method.

4.1. Experimental Setup for Measurement

To simplify the problem, consider that there are three coils mutually coupled with each other. Figure 5 shows the experiment setup to verify the performance of the proposed method. Three coils were aligned co-axially. The coupling coefficients between the first and third coils were measured by changing the distance between the two coils in the HF band, in which the second coil was always located in the middle of the two coils. The parameters of the coils are summarized in Table 1. The coil with the inductance of 1.1 µH was designed to have an high quality factor of about 250 at 6.78 MHz, and it has three turns and a single layer printed on the FR-4 substrate, of which the dimension is 95.7 mm × 105.7 mm. The coil of 3.8 µH was fabricated to verify the proposed method for coils with high inductance, and it has six turns and two layers printed on the FR-4 substrate of the same dimension as the coil of 1.1 µH.

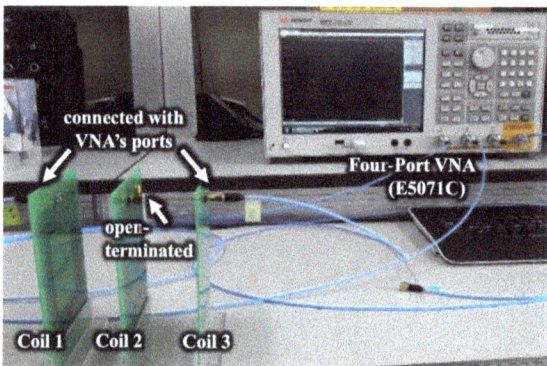

Figure 5. Experimental setup for measuring the coupling coefficient of three coils.

Table 1. Parameters of coils used in measurement.

	Coil 1	Coil 2	Coil 3
Case #1	$L = 1.1\ \mu H$ $R_{dc} = 0.08\ \Omega$	$L = 1.1\ \mu H$ $R_{dc} = 0.08\ \Omega$	$L = 1.1\ \mu H$ $R_{dc} = 0.08\ \Omega$
Case #2	$L = 1.1\ \mu H$ $R_{dc} = 0.08\ \Omega$	$L = 3.8\ \mu H$ $R_{dc} = 0.19\ \Omega$	$L = 1.1\ \mu H$ $R_{dc} = 0.08\ \Omega$

4.2. Changing Coupling Coefficient between Coils

Figure 6 shows the results measured by Method I using Keysight's four-port VNA, E5071C, for Cases #1, #2, and only two coils without a middle coil. Ideally, the coupling coefficient between two coils must maintain a constant value regardless of the frequency and existence of other coils. However, as the distance between the farthest coils decreases, the measured coupling coefficients deviate from that of the two coils without a middle coil. The deviation increases further as the frequency increases and the inductance of the middle coil increases.

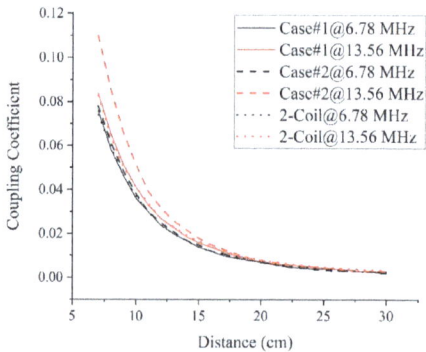

Figure 6. Measured coupling coefficient using Method I for three coils of Cases #1, #2, and two coils.

4.3. Comparison Results and Discussion

The coupling coefficients measured by the conventional and proposed methods with respect to the distance between the farthest coils are shown in Figure 7. We can notice that the proposed method produces almost the same results as the conventional methods. For a more accurate comparison, we calculated the mean and standard deviation of difference between the results of Method I and the proposed method, and the statistical results are summarized in Table 2. It is confirmed that both the mean and standard deviation are less than 0.001. This means that the proposed method provides precise and repeatable coupling coefficients among multiple coils.

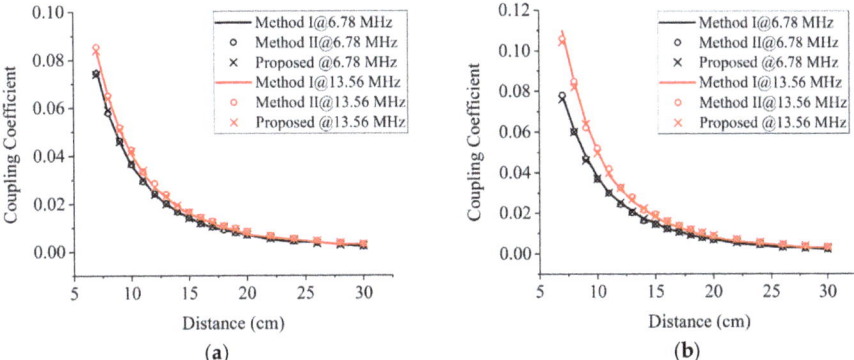

Figure 7. Measured coupling coefficients of three coils as a function of distance using the conventional and proposed methods at 6.78 and 13.56 MHz: (**a**) Case #1 and (**b**) Case #2.

Table 2. Mean and standard deviation of difference between the results of Method I and the proposed method.

	at 6.78 MHz		at 13.56 MHz	
	Mean	St Dev.	Mean	St Dev.
Case #1	4.1×10^{-4}	5.21×10^{-4}	5.11×10^{-4}	3.75×10^{-4}
Case #2	5.21×10^{-4}	5.94×10^{-4}	12.26×10^{-4}	11.73×10^{-4}

Figure 8 shows the reactance of transfer impedances measured by the conventional and proposed methods with respect to the frequency at the distance of 8 cm and 16 cm between the farthest coils. In spite of the simple measurement procedure, the proposed method provides accurate results that are similar to those of the conventional methods. That is, the proposed method can be used to measure the transfer impedance, as well as the coupling coefficient.

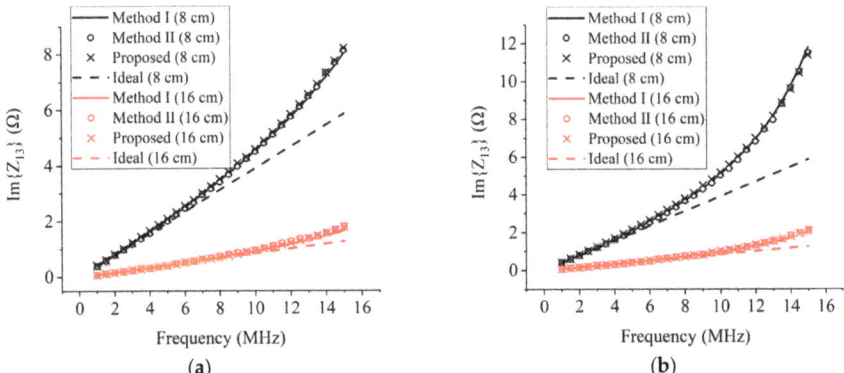

Figure 8. Measured transfer impedance Z_{13} as a function of frequency for the farthest distance of 8 cm and 16 cm using the conventional and proposed methods: (**a**) Case #1 and (**b**) Case #2.

Because the transfer impedance is ideally represented by $Z_{ij} = j\omega M_{ij}$, the transfer impedance should be proportional to the frequency. However, in reality, the transfer impedance is not linear for the frequency because of the effect of the parasitic components of the coil. These effects become stronger in Case #2 than Case #1 and at higher frequency than at lower frequency.

To validate the accuracy of the proposed procedure, the measured Z parameters were converted into the S parameters and they were compared with the results obtained from Methods I and II. First, in order to obtain a 3 × 3 [S] matrix from the proposed measurement method, a set of two coils was connected to a two-port VNA, and a 2 × 2 [Z] matrix was measured with the other coils open-circuited. For the other set of coils, 2 × 2 [Z] matrices were measured, and a 3 × 3 [Z] matrix was constructed by combining three 2 × 2 [Z] matrices. Finally, a 3 × 3 [S] matrix was obtained from a 3 × 3 [Z] matrix by applying conversion equations. Figure 9a,b shows the S_{13} from Methods I and II, and converted from the proposed method in a range from 1 to 15 MHz on the Smith chart and rectangular plot, respectively. S_{13} is the transmission coefficient from port 3 to port 1 when all other coils are terminated in matched load of 50 ohms. However, in reality, the coils of a WPT systems are not terminated with 50 ohms, and then S_{13} has no special physical meaning. Nevertheless, the [S] matrix converted from the [Z] matrix obtained by the proposed method is in good agreement with the [S] matrix obtained by the conventional methods in the frequency range from the LF to HF band. This means that the proposed method provides accurate measurement results.

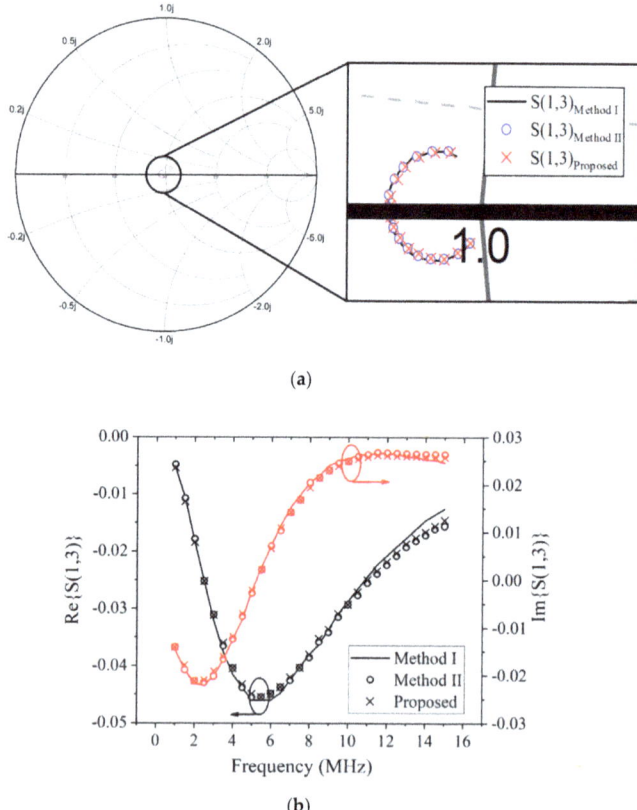

Figure 9. S parameter results for three coils from the proposed and conventional methods at 8 cm: (**a**) Smith chart form and (**b**) real and imaginary values.

In the proposed method, the critical assumption is that unconnected coils with a two-port network analyzer must be perfectly open-terminated. In the measurement, we realized the open condition by un-connecting anything from the coil's SMA connector. This method does not fully satisfy the assumption for wide frequency band or high frequency due to the parasitic capacitance of the SMA connector, whereas the un-connecting method can be still powerful at a low frequency, including LF and HF bands used in most WPT systems.

In order to investigate the effect of non-perfect open, we applied the proposed method to the measurement of the transfer impedance of three coils with open-termination and without any termination. The used open termination is the OPEN part of the Keysight's 85052C, of which the error is less than ±0.65° for DC to 3 GHz. Figure 10 shows the normalized coupling coefficients with respect to frequencies that were obtained from the measured transfer impedance using Equation (2) and normalized to the maximum coupling coefficient for comparison of the effect of the open termination. The difference between the coupling coefficients of the two cases is within approximately 5%, except for around 20 MHz and 50 MHz, which are the self-resonant frequencies of the coils of 3.8 µH and 1.1 µH, respectively. In other words, this means that the proposed method provides reliable results without ideal open termination, but must not be used near the SRF of the coils. Since it is common to operate WPT systems at a lower frequency than the SRFs of coils, the proposed method can be effectively applied to measure the coupling coefficient or mutual inductance between the coils used in WPT systems.

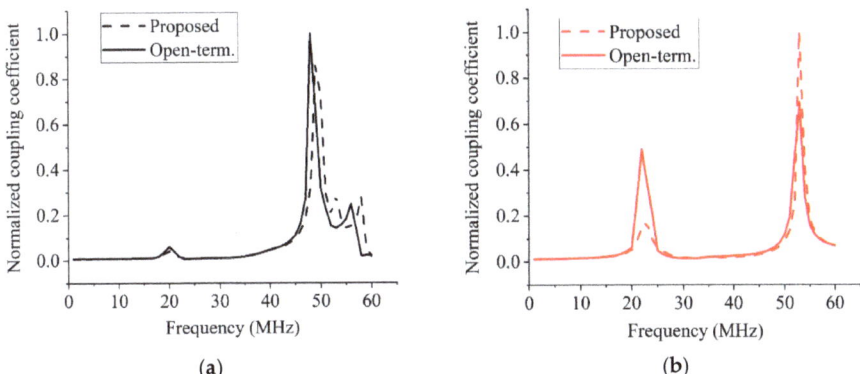

Figure 10. Measured normalized coupling coefficient at distance of 8 cm for open-terminated methods: (**a**) Case #1 and (**b**) Case #2.

5. Conclusions

In this paper, a coupling coefficient measurement method with a simple procedure was proposed to overcome the inconvenience and limitation of conventional measurement methods of coupling coefficients among coils in multi-coil WPT systems. To verify the potential of the proposed measurement method, a comparative analysis was performed by graph and mean and standard deviation values with the conventional measurement methods with respect to distance. As a result, the proposed method was demonstrated to achieve competitive performance with good accuracy. If the proposed method is applied to the magnetic beamforming of a multi-coil WPT systems, it can be a very powerful alternative to obtain prompt results.

Author Contributions: Conceptualization, D.-W.S.; methodology, D.-W.S.; software, S.-J.J.; validation, S.-J.J.; formal analysis, S.-J.J.; investigation, S.-J.J.; resources, S.-J.J.; data curation, D.-W.S.; writing—original draft preparation, D.-W.S.; writing—review and editing, D.-W.S.; visualization, S.-J.J.; supervision, D.-W.S.; project administration, D.-W.S.; funding acquisition, D.-W.S.

Funding: This work was supported by a National Research Foundation of Korea (NRF) grant funded by the Korea Government (MSIT) (No. 2018R1C1B6003854).

Conflicts of Interest: The authors declare no conflicts of interest.

References

1. Kalwar, K.A.; Aamir, M.; Mekhilef, S. A design method for developing a high misalignment tolerant wireless charging system for electric vehicles. *Measurement* **2018**, *118*, 237–245. [CrossRef]
2. Mao, S.; Wang, H.; Zhu, C.; Mao, Z.-H.; Sun, M. Simultaneous wireless power transfer and data communication using synchronous pulse-controlled load modulation. *Measurement* **2017**, *109*, 316–325. [CrossRef] [PubMed]
3. Anowar, T.I.; Barman, S.D.; Reza, A.W.; Kumar, N. High-Efficiency Resonant Coupled Wireless Power Transfer via Tunable Impedance Matching. *Int. J. Electron.* **2017**, *104*, 1607–1625.
4. Lee, J.H.; Seo, D.W. Adaptive Wireless Power Transfer System without Feedback Information using Single Matching Network. *IEICE Trans. Commun.* **2019**, *E102-B*, 257–265. [CrossRef]
5. Jadidian, J.; Katabi, D. Magnetic MIMO: How To Charge Your Phone in Your Pocket. In Proceedings of the 20th Annual International Conference on Mobile Computing and Networking (MOBICOM), Maui, HI, USA, 7–11 September 2014; pp. 495–506.
6. Shi, L.; Kabelac, Z.; Katabi, D.; Perreault, D. Wireless Power Hotspot that Charges All of Your Devices. In Proceedings of the 21th Annual International Conference on Mobile Computing and Networking (MOBICOM), Paris, France, 7–11 September 2015; pp. 2–13.
7. Acero, J.; Carretero, C.; Lope, I.; Alonso, R.; Lucia, O.; Burdio, J.M. Analysis of the mutual inductance of planar-lumped inductive power transfer systems. *IEEE Trans. Ind. Electron.* **2013**, *60*, 410–420.
8. Theodoulidis, T.; Ditchburn, R.J. Mutual impedance of cylindrical coils at an arbitrary position and orientation above a planar conductor. *IEEE Trans. Magn.* **2007**, *43*, 3368–3370. [CrossRef]
9. Fu, M.; Zhang, T.; Ma, C.; Zhu, X. Efficiency and optimal loads analysis for multiple-receiver wireless power transfer systems. *IEEE Trans. Microw. Theory Tech.* **2015**, *63*, 801–812.
10. Hackl, S.; Lanschutzer, C.; Raggam, P.; Randeu, W.L. A Novel Method for Determining the Mutual Inductance for 13.56 MHz RFID Systems. In Proceedings of the 2008 6th International Symposium on Communication Systems, Networks and Digital Signal Processing, Graz, Austria, 25 July 2008; pp. 297–300.
11. Tippet, J.C.; Speciale, R.A. A rigorous technique for measuring the scattering matrix of a multiport device with a 2-port network analyzer. *IEEE Trans. Microw. Theory Tech.* **1982**, *30*, 661–667. [CrossRef]
12. Rolfes, I.; Schiek, B. Multiport method for the measurement of the scattering parameters of n-ports. *IEEE Trans. Microw. Theory Tech.* **2005**, *53*, 1990–1996. [CrossRef]

© 2019 by the authors. Licensee MDPI, Basel, Switzerland. This article is an open access article distributed under the terms and conditions of the Creative Commons Attribution (CC BY) license (http://creativecommons.org/licenses/by/4.0/).

Article

Numerical and Circuit Modeling of the Low-Power Periodic WPT Systems

Adam Steckiewicz *, Jacek Maciej Stankiewicz and Agnieszka Choroszucho

Faculty of Electrical Engineering, Bialystok University of Technology, Wiejska 45D, 15-351 Bialystok, Poland; j.stankiewicz@doktoranci.pb.edu.pl (J.M.S.); a.choroszucho@pb.edu.pl (A.C.)
* Correspondence: a.steckiewicz@pb.edu.pl

Received: 5 April 2020; Accepted: 19 May 2020; Published: 22 May 2020

Abstract: This article presents a method for analysis of the low-power periodic Wireless Power Transfer (WPT) system, using field and circuit models. A three-dimensional numerical model of multi-segment charging system, with periodic boundary conditions and current sheet approximation was solved by using the finite element method (FEM) and discussed. An equivalent circuit model of periodic WPT system was proposed, and required lumped parameters were obtained, utilizing analytical formulae. Mathematical formulations were complemented by analysis of some geometrical variants, where transmitting and receiving coils with different sizes and numbers of turns were considered. The results indicated that the proposed circuit model was able to achieve similar accuracy as the numerical model. However, the complexity of model and analysis were significantly reduced.

Keywords: wireless power transfer; wireless charging; circuit modeling; numerical analysis

1. Introduction

In the present days, we have observed a growing number of devices operating due to wireless power transfer (WPT) technology [1], which became more available in extensive scattered grids of many interdependent sources and loads [2]. Current trends in wireless charging of electric vehicles [3,4] and modern electronics [1,5,6] have led to the development of the inductive power transfer (IPT) concept. Among other things, an increasing number of mobile devices processing huge amounts of data [7,8] is directly connected with their computing power and number of sensors. Nowadays, WPT is considered to be an alternative method of charging wireless devices, where a pair of coils [7,9] (accompanied with additional intermediate coils [10,11]) or an array of coils [12–14] is utilized. Multi-coil systems operate at high frequencies ($f \geq 1$ MHz) [13,15], and in some cases, power transfer is assisted by using metamaterial structures [14]. For low frequencies ($f < 1$ MHz), an array of coils as domino form resonators [16] and linear resonator arrays [17,18] are considered, where in intermediate space between transmitter and receiver, energy transfer is assisted by using several resonators. However, a detailed analysis was performed for a series configuration of resonators, while parallel-series topology of planar coils, acting as group of energy transmitters and receivers, are still not fully developed. Wireless charging is also considered in the systems of beacons [19] in hard-to-reach places, medical implants in human body [20], and smart buildings with sensors inside rooftops and walls [21].

Energy supply or charging of many devices located in close range to each other may be simplified by using WPT systems as a grid of periodically arranged coils which forms surfaces for transmitting or receiving the energy. This solution increases the density of transferred power, and also simultaneous energy supply (using single power source) for many devices is possible. Potential applications of this system are mainly focused on the simultaneous charging of an array of sensors (embedded in, e.g., walls or floors) and sets of implantable electronic devices placed inside the body [22]. From the point of view of high-power applications, proposed models of periodic WPT surfaces may be utilized as an analysis method when charging vehicles on large parking spaces is considered.

This article presents a wireless power transfer system with periodically arranged planar coils. The main purpose of this work is to introduce and study numerical and circuit model, which can be applied to analyze power transfer conditions in discussed systems. Both approaches reduce size and complexity of typically utilized numerical and circuit models. The proposed unit cell analysis with periodic boundary conditions does not require a full 3D model with many coils [23] in which the number of degrees of freedom is significant. A simplified model in the form of a well-known T-type equivalent circuit is an alternative for more extensive matrix formulation [11,16,17], where a large coefficient matrix with lumped parameters has to be known. Both models make it possible to evaluate the influence of the coil structure on power transfer. Adjusting the geometrical parameters gives an ability to obtain high efficiency of the power transfer to multiple loads. A numerical analysis of the time-harmonic magnetic field in a 3D model of the system is characterized, and, on this basis, the efficiency and power transfer conditions are specified. The simplified circuit model is proposed, and the required lumped parameters are calculated by using analytical formulae. The computational results in the frequency domain of the exemplary periodic WPT systems, performed in numerical software, are compared with the results obtained from an equivalent circuit. The authors analyzed the influence of geometrical parameters (coil radius, number of turns, and distance between coils) on power transfer efficiency, as well as transmitter and receiver currents.

2. Materials and Methods

2.1. Periodic Wireless Power Transfer System

Among typical WPT devices consisting of several coils, systems with many inductive elements may also be considered. A pair of transmitting (TR) and receiving (RE) circular inductors at the distance, h, possessing identical radius, r_c, and number of turns, n_c, are the fundamental parts of the WPT cell with outer dimensions $d_c \times d_c$ (Figure 1). Windings are wound around a dielectric carcass with additional compensating capacitors. The periodic distribution of WPT cells (Figure 1) leads to transmitting and receiving surfaces where the energy transmission occurs. The transmitting surface consists of TR coils connected parallel to the sinusoidal voltage source (RMS value U_t), while RE coils are connected with individual loads, $\underline{Z}_J$.

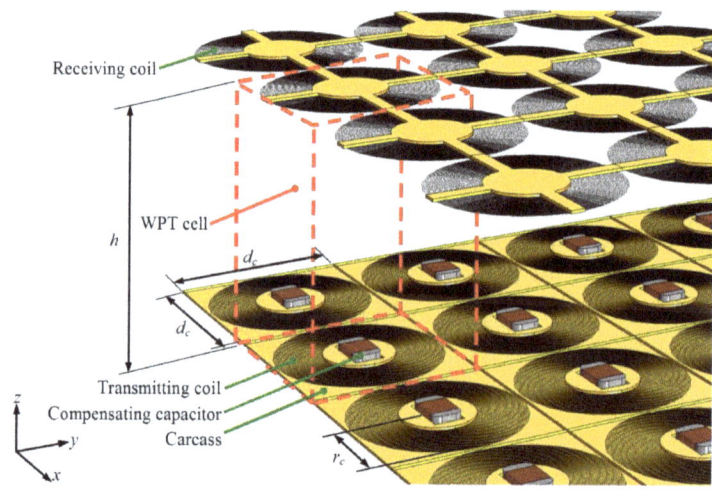

Figure 1. Periodic WPT system combined with an array of WPT cells.

Proposed configuration enhances the density of transferred power in an area between transmitting and receiving surfaces. Furthermore, the energy supply conditions can be adjusted. For example, the simultaneous power transfer to many independent devices is possible, where each WPT cell is directly connected with individual energy storage. Another possibility is to connect parallel every RE coil to a single common energy receiver. The series connection of coils and intermediary parallel-series configuration are possible. An analysis of the periodic system can be reduced to the two-dimensional plane, xy (Figure 2), representing a set of TR or RE coils. The considered cell, $\Theta_{x+a,y+b}$, is an element of an array with identical inductors, where a is the number of columns and b is the number of rows in a grid; $a, b \in \mathbf{Z}$, and $\mathbf{Z}$ are the set of integers. Adjacent coils (e.g., $\Theta_{x,y+1}$ or $\Theta_{x-1,y}$) of element $\Theta_{x,y}$ are separated by the distance, d_c. Magnetic coupling, which occurs between coil $\Theta_{x,y}$ and the others, is undesirable and affects power transfer efficiency between transmitting and receiving surfaces. Due to the small distance between coils ($d_c \approx 2r_c$), magnetic coupling phenomena must be included in models.

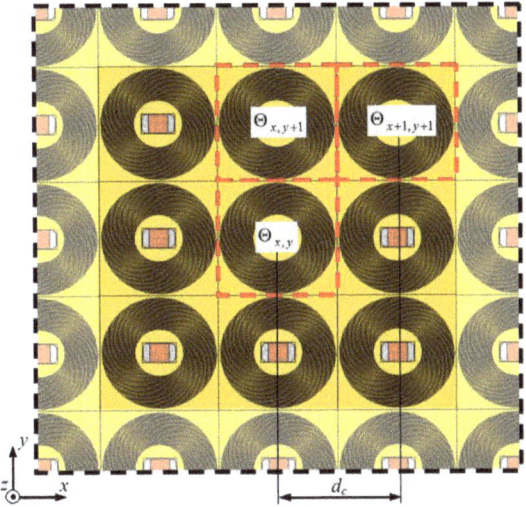

Figure 2. Transmitting/receiving surface of the periodic WPT system: $\Theta_{x,y}$—WPT cell, $\Theta_{x,y+1}$—adjacent WPT cell (by edge), $\Theta_{x+1,y+1}$—adjacent WPT cell (by vertex).

2.2. Modeling Approach

The analysis of a periodic wireless charging system may be performed by using numerical methods or experimental research of some prototypes. An application of simulation software gives an ability to create a numerical model of the system and to find a distribution of magnetic field. However, a three-dimensional model is required, as well as complex boundary conditions. Effectiveness and accuracy of the obtained solution arise from model size (number of degrees of freedom, NDOF). A greater number of degrees of freedom results in greater accuracy of solution but also leads to a longer calculation time. On the other hand, during the experimental research, it is necessary to build several prototypes with many coils and specified geometry. While it is possible to examine the impact of electrical parameters (e.g., current frequency and load impedance) on wireless power transfer, the potential identification of geometrical parameters (e.g., coil radius and number of turns) is limited.

At the design stage and initial analysis of periodic WPT charging system and its properties (e.g., efficiency, power losses, and load power), mathematical models are sufficient. Hence, two possible approaches were characterized:

- Numerical model of periodic WPT system, with necessary simplifications and boundary conditions.
- Circuit model as an alternative for numerical model.

The usage of electrical circuit helps to avoid the numerical analysis and building a series of prototypes subjected to experiments.

2.3. Numerical Model

A numerical analysis of energy transfer in the system combined with many WPT cells requires taking into account many details of the model, such as the following:

- Coil geometry,
- Winding structure, number of WPT cells,
- Electrical elements (e.g., compensating capacitors, loads) connected to coils.

Planar spiral coils were wound of several dozens of turns, made of ultra-thin wires with diameter d_w. In order to reduce NDOF, current sheet approximation [24–26] was applied, which replaces the multi-turn coil with a homogeneous structure (Figure 3). Current sheet is a model for a group of wires wound together around a specified carcass, but still insulated from each other by an electrical insulator of a thickness d_i. The current flows in the direction of wires (*xy* plane), while current densities in other directions are omitted. To correctly apply this method of approximation, one may make the following assumptions:

$$n_c \geq 10, \tag{1a}$$

$$d_w < \delta, \tag{1b}$$

$$d_i \ll d_w, \tag{1c}$$

where n_c is the number of turns, δ is the penetration depth, and d_i is the wire insulation thickness. Without current sheet approximation, Assumptions (1a) and (1c) impose the necessity to include every turn. As a consequence, this increases NDOF, which makes the numerical model difficult to solve using typical computational units.

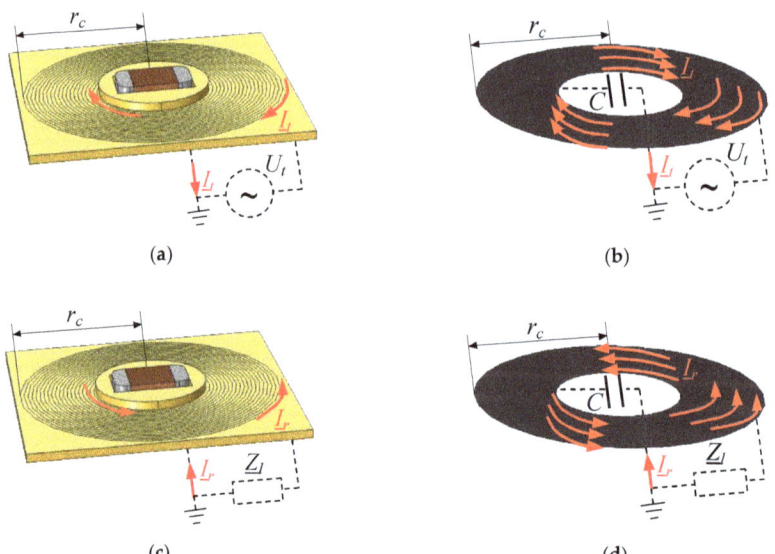

Figure 3. Models of multi-turn spiral coils: full model of (**a**) transmitting coil and (**c**) receiving coil; simplified model, using current sheet approximation method and attached electrical circuit for (**b**) transmitting coil and (**d**) receiving coil.

Compensating capacitor can be modeled as an element with lumped capacity, C. Additionally, it is possible to omit a carcass if it is made of dielectric and non-magnetic material ($\mu = \mu_0$). A voltage source with RMS value U_t and frequency f is connected to each coil and current $\underline{I}_t$ flows through transmitter. Receiving coil, connected with a linear load, $\underline{Z}_l$, carry induced current $\underline{I}_r$.

In order to simulate the periodic WPT system (Figure 1), all the cells forming transmitting and receiving surfaces have to be taken into account. However, for the system with many WPT cells ($a, b \gg 3$), another simplification is possible. Assuming $a, b \rightarrow \pm\infty$ periodic boundary conditions (PC) both in x and y direction may be applied. Then, wireless charging system will be simplified to a single cell $\Theta_{x,y}$, filled with air and containing a pair of transmitting and receiving coils (Figure 4). Periodic boundary conditions are applied on the left and right (PC_x), as well as the front and back (PC_y) boundaries, in order to project an infinite array of WPT cells. A perfectly matched layer (PML) is put at the top and bottom of the model, to imitate a dielectric background. The model is complemented by application of simplified multi-turn spiral coils with an attached part of the electrical circuit, as shown in Figure 3b,d.

The energy transport problem in the presented system (Figure 4) can be solved by using magnetic vector potential $\mathbf{A} = [A_x \ A_y \ A_z]$ and formulation of magnetic field phenomena in frequency domain, using the Helmholtz equation:

$$\nabla \times \left(\mu_0^{-1} \nabla \times \mathbf{A}\right) - j\omega\sigma\mathbf{A} = \mathbf{J}_{ext}, \tag{2}$$

where μ_0 is the permeability of air (H/m), ω is the angular frequency (rad/s), σ is the electrical conductivity (S/m), and $\mathbf{J}_{ext}$ is the external current density (A/m^2). Periodic boundary conditions on four external surfaces were defined as a magnetic insulation:

$$\mathbf{n} \times \mathbf{A} = 0, \tag{3}$$

where $\mathbf{n} = [1_x \ 1_y \ 1_z]$ is a surface normal vector. Voltage supply (U_t) has direct impact on $\mathbf{J}_{ext}$, and when combined with Equation (3), it enables us to solve Equation (2) by using numerical methods, e.g., finite element method (FEM). Then, the volume distribution of vector potential $\mathbf{A}(x,y,z)$ can be found. The capacity of the compensating capacitor may be defined from the parametric analysis of the system for different C. When Im[$\underline{I}_t$] $\approx$ 0 one may assume, that the resonant state was reached and adjusted value of C is a required capacity.

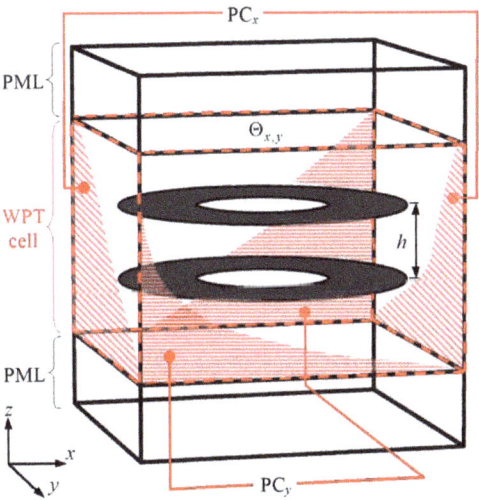

Figure 4. Numerical model of the periodic WPT system.

2.4. Circuit Model

The formulation and solution of a numerical model of a periodic WPT system is a multi-task problem, requiring advanced numerical methods. Despite an ability of performing simulation on typical computational units, it is desirable to propose a simpler model that still will be able to ensure similar analysis, but faster modeling and less-complex calculations. As an alternative, we proposed a circuit model (Figure 5) combining two-port network with analytical formulae for calculating lumped parameters. Similar to the numerical model, the infinite periodic grid would be simplified to analysis of a single WPT cell. The solution of the circuit model in the frequency domain can be performed by using methods of circuit analysis; however, the main issue is to determine the values of several lumped parameters. It is necessary to take into account the impact of adjacent cells on inductances L_t and L_r of TR coil and RE coil, as well as their mutual inductance, M_{tr}.

Resistance of a coil may be found by replacing spiral structure of windings, using concentering circles possessing identical widths, $d_w + d_i$ (Figure 6). Starting from the outer edge, the mean length of each circle is described by the following:

$$l_n = \pi[2r_c - (2n-1)(d_w + d_i)], \tag{4}$$

Hence, total length of all circles is defined as follows:

$$l_c = \sum_{n=1}^{n_c} l_n = \pi n_c [2r_c - n_c(d_w + d_i)]. \tag{5}$$

By substituting Equation (5) to the formula determining resistance of a conductor with constant cross section, resistance of an inductor can be found:

$$R_c = \frac{l_c}{\sigma \pi \left(\frac{d_w}{2}\right)^2} = \frac{4n_c[2r_c - n_c(d_w + d_i)]}{\sigma d_w^2}. \tag{6}$$

If coils (TR and RE) are identical and the considered frequency bandwidth condition (1b) is met, calculated resistances $R_t = R_r = R_c$ will not be dependent of frequency.

Self-inductance of a spiral planar coil can be calculated by using the following formula [27]:

$$L_{self} = \frac{1}{2} c_1 \mu_0 d_{avg} n_c^2 \left[\ln\left(\frac{c_2}{\rho}\right) + c_3 \rho + c_4 \rho^2 \right], \tag{7}$$

where d_{avg} is a mean diameter

$$d_{avg} = 2r_c - (d_w + d_i) n_c, \tag{8}$$

and ρ is a fill factor

$$\rho = \frac{(d_w + d_i) n_c}{2r_c - (d_w + d_i) n_c}, \tag{9}$$

while coefficients $c_1, c_2, c_3,$ and c_4 are depending on geometry (shape) of a coil [27]. For identical TR and RE coils calculated inductances are equal, $L_t = L_r = L_c$ (Figure 5).

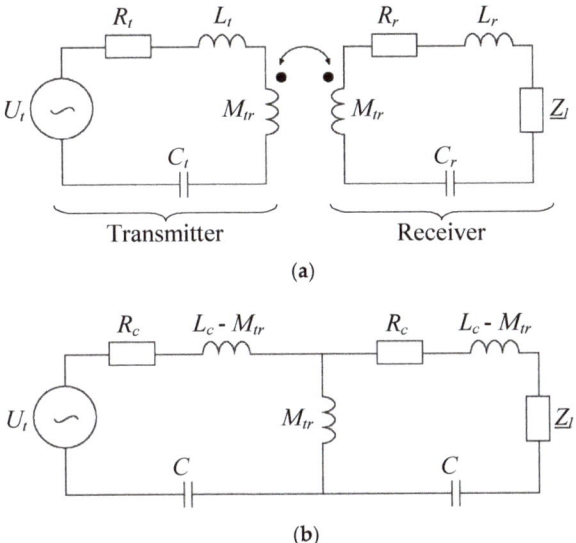

Figure 5. Circuit model of the WPT cell with identical transmitting and receiving coils: (**a**) general model of periodic cell and (**b**) simplified model of the cell for identical transmitting and receiving coil.

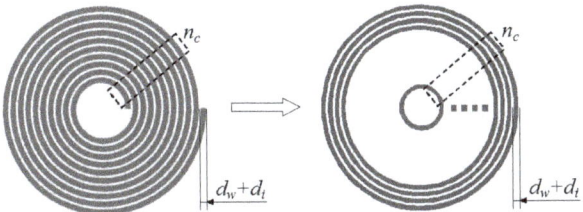

Figure 6. Spiral coil approximation for resistance calculation, using concentering circles.

In the periodic grid, coils are adjacent; hence, it is necessary to include magnetic coupling between them. Mutual inductance M_{period}, which came directly from periodic distribution of coils arranged on the surface xy, is a sum of all mutual inductances [28,29]:

$$M_{period} = \sum_a \sum_b \left(M_{x+a,y+b}\right) - M_{x,y}, \tag{10}$$

where $M_{x+a,y+b}$ is the mutual inductance between coil at coordinate (x,y) and coil at a-th column and b-th row; $M_{x,y} = L_{self}$ is self-inductance. The following assumptions are then taken into account:

- Only coupling between adjacent coils is considered ($|a|_{max} = |b|_{max} = 1$),
- The system is periodic and symmetrical ($M_{x+a,y+b} = M_{x-a,y-b}$),
- Mutual inductances of coils adjacent to $\Theta_{x,y}$ are assumed to be approximately equal ($M_{x+a,y} \approx M_{x,y+b} \approx M_{x+a,y+b}$),

By taking into account the above assumptions, Equation (10) can be simplified as follows:

$$M_{period} = 8 M_{x,y+1}, \tag{11}$$

where $M_{x,y+1}$ is the mutual inductance between coil at coordinate (x,y) and an edge adjacent coil (Figure 2). For calculation of $M_{x,y+1}$, the formula introduced by Siu, Su, and Lai [30] is suitable:

$$M_{x,y+1} = \frac{\mu_0 g^2}{4\pi} \int_{\Phi_i}^{\Phi_o} \int_{\Phi_i}^{\Phi_o} \frac{[(1+\varphi_1\varphi_2)\cos(\varphi_2-\varphi_1) - (\varphi_2-\varphi_1)\sin(\varphi_2-\varphi_1)]d\varphi_1 d\varphi_2}{\sqrt{(d_c + g\varphi_2\cos\varphi_2 - g\varphi_1\cos\varphi_1)^2 + (g\varphi_2\sin\varphi_2 - g\varphi_1\sin\varphi_1)^2}}, \quad (12)$$

where $g = (d_w + d_i)/(2\pi)$, $\Phi_i = [r_c - (d_w + d_i)n_c]/g$, $\Phi_o = r_c/g$. In the literature, no analytical solution for Equation (12) was found; however, it is possible to find it by using numerical integration. After applying the rectangle rule formula, Equation (12) takes the following form:

$$M_{x,y+1} = \frac{\mu_0 g \Phi_K}{4\pi} \sum_{k_2=1}^{K} \sum_{k_1=1}^{K} \frac{\left(1 + k_1 k_2 \Phi_K^2\right)\cos(k_2\Phi_K - k_1\Phi_K) - (k_2\Phi_K - k_1\Phi_K)\sin(k_2\Phi_K - k_1\Phi_K)}{\sqrt{\left(\frac{d_c}{g\Phi_K} + k_2\cos k_2\Phi_K - k_1\cos k_1\Phi_K\right)^2 + (k_2\sin k_2\Phi_K - k_1\sin k_1\Phi_K)^2}}, \quad (13)$$

where $\Phi_K = (\Phi_o - \Phi_i)/K$ is an integration step, while K is assumed number of integration subintervals, $K \geq r_c/g$ and $K \in \mathbb{N}$.

Horizontal periodicity affects the magnetic field of an arbitrary coil, where the opposite magnetic field of neighboring inductors reduces total magnetic energy associated with this coil. As a consequence, its effective inductance, L_c, will be less than self-inductance, L_{self}. For the total mutual inductance of Equation (11), the effective inductance of the considered coil in segment $\Theta_{x,y}$ will be defined as follows:

$$L_c = L_{self} + M_{period} = L_{self} + 8M_{x,y+1}, \quad (14)$$

In the next step, after calculations of self-inductance, L_{self}, using Equations (7)–(9) and total mutual inductance in periodic grid $M_{x,y+1}$ from Equation (13), both quantities are substituted to Equation (14), in order to find effective inductance L_c. On the basis of a series resonant and known value of L_c, it is possible to find the compensating capacity, C, at a specified frequency.

$$C(f) = \frac{1}{4\pi^2 f^2 L_c} = \frac{1}{4\pi^2 f^2 (L_{self} + M_{period})} = \frac{1}{4\pi^2 f^2 (L_{self} + 8M_{x,y+1})}, \quad (15)$$

where $C_t = C_r = C(f)$, if it was assumed that TR and RE coils are identical.

Mutual inductance M_{tr} may be presented in the following form:

$$M_{tr} = k_p M_z, \quad (16)$$

where mutual inductance M_z between transmitter and receiver is calculated from the following [30]:

$$M_z = \frac{\mu_0 g^2}{4\pi} \int_{\Phi_i}^{\Phi_o} \int_{\Phi_i}^{\Phi_o} \frac{[(1+\varphi_1\varphi_2)\cos(\varphi_2-\varphi_1) - (\varphi_2-\varphi_1)\sin(\varphi_2-\varphi_1)]d\varphi_1 d\varphi_2}{\sqrt{h^2 + g^2\varphi_1^2 + g^2\varphi_2^2 - 2g^2\varphi_1\varphi_2\cos(\varphi_2-\varphi_1)}}, \quad (17)$$

and after an application of rectangle rule, Equation (17) has the following form:

$$M_z = \frac{\mu_0 g^2 \Phi_K^2}{4\pi} \sum_{k_2=1}^{K} \sum_{k_1=1}^{K} \frac{\left(1 + k_1 k_2 \Phi_K^2\right)\cos(k_2\Phi_K - k_1\Phi_K) - (k_2\Phi_K - k_1\Phi_K)\sin(k_2\Phi_K - k_1\Phi_K)}{\sqrt{h^2 + g^2(k_1\Phi_K)^2 + g^2(k_2\Phi_K)^2 - 2g^2 k_1 k_2 \Phi_K^2 \cos(k_2\Phi_K - k_1\Phi_K)}}. \quad (18)$$

Periodic coupling coefficient, k_p, results from physical phenomena in which the magnetic field of all coils in the system affects mutual inductance, M_z. As a result, $M_{tr} < M_z$, which means that, for periodic WPT, M_{tr} between TR and RE is reduced by some factor k_p. In other words, the k_p is related to magnetic couplings between coils adjacent to $\Theta_{x,y}$ (reducing L_{self} by M_{period}), as well as to power transfer between neighboring WPT cells. If numerical or experimental data for particular systems are known, it is possible to find k_p by comparing these data with those obtained from an equivalent circuit.

A different way is analytical derivation of coupling coefficient, which is a very complex task. Therefore, an empirical formula was proposed as a simplification for presented small-scale systems:

$$k_p = \exp(-\lambda \cdot h / r_c),\qquad(19)$$

where λ is an approximation function coefficient. Based on a set of numerical results (Figure 7) for different h/r_c, the authors have derived $\lambda = 1.2252$ as an optimal value for exponential approximation function (19). Then, substituting parameters calculated from Equations (18) and (19) to (16), it is possible to find mutual inductance, M_{tr}, for the WPT cell, which is applicable at, for example, the early design stage.

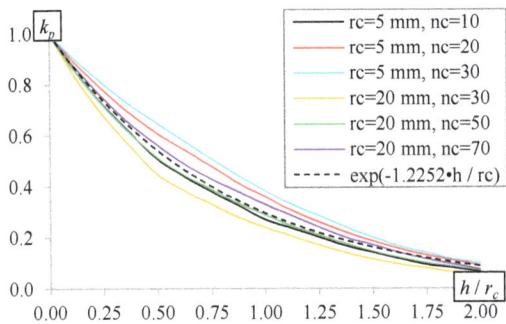

Figure 7. Periodic coupling coefficient, k_p, for considered coils and different h/r_c.

3. Results and Discussion

3.1. Analyzed Models

The numerical field model took into account the electromagnetic phenomena and geometrical structure of the WPT cell; hence, it was a reference for the simplified circuit model. On the basis of obtained results for several exemplary periodic WTP system, the authors have verified the validity of its electrical model by comparing absolute current of TR (I_t) and RE (I_r) coils, as well as energy transfer efficiency, η. Since passive load, Z_l, was considered, its active power was calculated by using the following formula:

$$P_l = Z_l I_r^2.\qquad(20)$$

Because of the resonant state obtained after an application of the compensating capacitor, the imaginary part of the transmitter current was negligible ($\text{Im}[\underline{I}_t] \approx 0$); hence, voltage source produced only active power.

$$P_s = U_t I_t.\qquad(21)$$

Finally, using Equations (20) and (21), we found the power transfer efficiency:

$$\eta = \frac{P_l}{P_s} 100\%.\qquad(22)$$

In the further part of this section about the characteristics of I_t, I_r, and η with the label FM (field model) were related to numerical model and with the label EC to electrical circuit.

We subjected to analysis discussed unit cell $\Theta_{x,y}$, where we assumed that the system consists of an infinite number of WPT cells. Every cell consisted of a pair of identical coaxial coils arranged at a distance, h, and wounded using wire with a diameter of $d_w = 150$ μm, insulation thickness $d_i = 1$ μm, and conductivity $\sigma = 5.6 \cdot 10^7$ S/m. When L_{self} was calculated by using Equation (7), we assumed $c_1 = 1$, $c_2 = 2.5$, $c_3 = 0$, and $c_4 = 0.2$. Voltage supply with RMS value $U_t = 5$ V and frequency from $f_{min} = 0.1$ MHz to $f_{max} = 1$ MHz was attached to the TR coil. Passive load $Z_l = 50$ Ω was connected with the RE coil. We analyzed small- ($r_c = 5$ mm) and large-size coils ($r_c = 20$ mm) with a different number of turns, n_c, distance, h (Table 1), and constant separation between neighboring cells, $d_c = 2.25 r_c$. The numerical model (Figure 4) created in *Comsol Multiphysics* software was solved by FEM. We utilized built-in multi-turn coils' approximation and partial electrical circuit combined with a 3D model. Lumped parameters of electrical circuit of Figure 5b (Table 2) were found by using Equations (6), (7), (13), (15), (18), and (19). Transmitter and receiver currents, as well as power transfer efficiency (Equation (22)), were calculated for both models, within frequency range $f_{min} \div f_{max}$.

Table 1. Geometrical parameters for considered cases.

r_c (mm)	n_c	h (mm)		
		$0.5 r_c$	r_c	$2 r_c$
5	10	2.5	5.0	10.0
	20	2.5	5.0	10.0
	30	2.5	5.0	10.0
20	30	10.0	20.0	40.0
	50	10.0	20.0	40.0
	70	10.0	20.0	40.0

Table 2. Lumped parameters of the electrical circuit.

r_c (mm)	n_c	R_c (Ω)	L_{self} (H)	M_{period} (H)	C_c (F)	$h = 0.5 r_c$		$h = r_c$		$h = 2 r_c$	
						M_z (H)	k_p	M_z (H)	k_p	M_z (H)	k_p
5	10	0.274	1.41×10^{-6}	3.78×10^{-8}	2.28×10^{-8}	3.68×10^{-7}	0.542	1.54×10^{-7}	0.293	4.06×10^{-8}	0.086
	20	0.453	3.14×10^{-6}	6.97×10^{-8}	9.80×10^{-9}	8.96×10^{-7}	0.542	3.56×10^{-7}	0.293	8.74×10^{-8}	0.086
	30	0.535	3.84×10^{-6}	7.74×10^{-8}	7.86×10^{-9}	1.08×10^{-6}	0.542	4.21×10^{-7}	0.293	1.01×10^{-7}	0.086
20	30	3.393	5.97×10^{-5}	1.62×10^{-6}	5.41×10^{-10}	1.53×10^{-5}	0.542	6.47×10^{-6}	0.293	1.73×10^{-6}	0.086
	50	5.175	1.22×10^{-4}	3.00×10^{-6}	2.60×10^{-10}	3.45×10^{-5}	0.542	1.41×10^{-5}	0.293	3.60×10^{-6}	0.086
	70	6.574	1.78×10^{-4}	3.99×10^{-6}	1.73×10^{-10}	5.27×10^{-5}	0.542	2.10×10^{-5}	0.293	5.17×10^{-6}	0.086

3.2. Model Comparison and Electrical Parameters

At the beginning, computations of small-size coils ($r_c = 5$ mm) were performed. The results from numerical and circuit model for $n_c = 10$ (Figure 8) were in a good agreement, since characteristics for different distances, h, and frequencies overlapped. However, WPT efficiency was below 10% (Figure 8c), even when TR and RE coils were close to each other ($h = 2.5$ mm)—in those cases, the number of turns was insufficient.

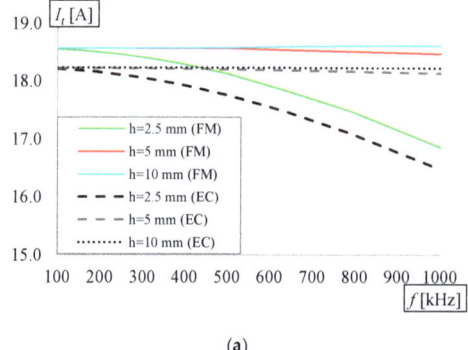

(a)

Figure 8. Cont.

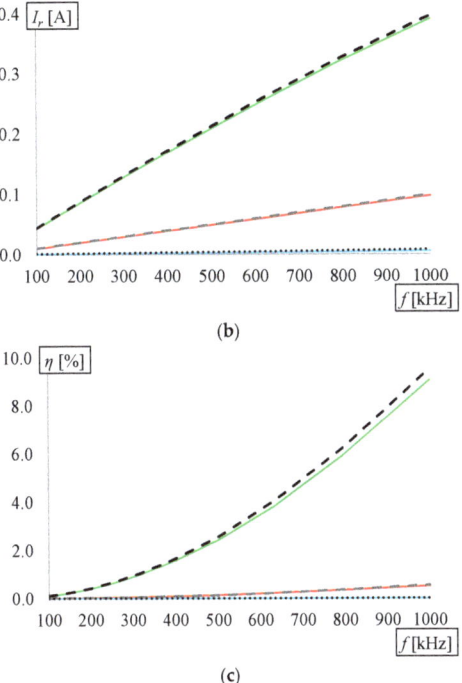

Figure 8. Results for the case r_c = 5 mm, n_c = 10: (**a**) transmitter current, (**b**) receiver current, and (**c**) power transfer efficiency.

The highest differences between FM and EC, especially those related to power transfer efficiency, η, were observed at $h = r_c$ = 5 mm (Figure 9). Nonetheless, very good qualitative agreement for the entire bandwidth and all distances, h, was preserved. The increased number of turns resulted in higher efficiency (almost 40% at f = 1 MHz) and lower values of I_t with relation to the previous case. Still, negligible efficiency was achieved at $h = 2r_c$ = 10 mm, despite its increase with increasing frequency.

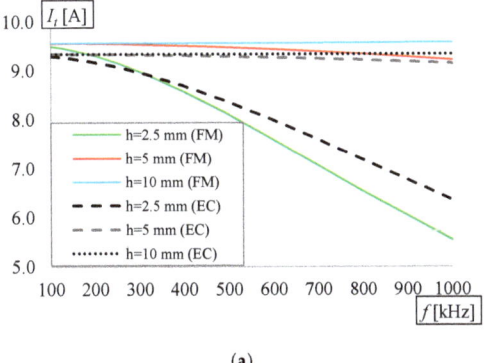

(**a**)

Figure 9. *Cont.*

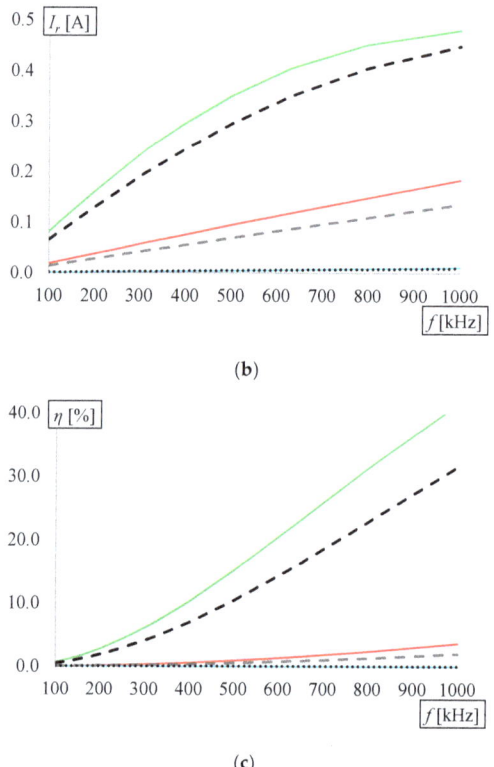

(b)

(c)

Figure 9. Results for the case $r_c = 5$ mm, $n_c = 30$: (**a**) transmitter current, (**b**) receiver current, and (**c**) power transfer efficiency.

For the larger coil ($r_c = 20$ mm), the shape of characteristics at $h = 0.5r_c = 10$ mm had changed (Figure 10). By comparing results at $n_c = 30$ for small and large coils, it was observed that I_t and I_r, as well as directly related source and load power, decreased significantly. The circuit model was able to follow that specific change in currents and efficiency characteristics, and a frequency range (approximately 200 ÷ 400 kHz) of the highest transmitted power (Figure 10b) was properly modeled.

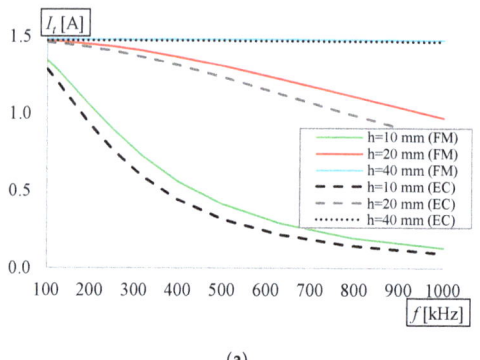

(a)

Figure 10. *Cont.*

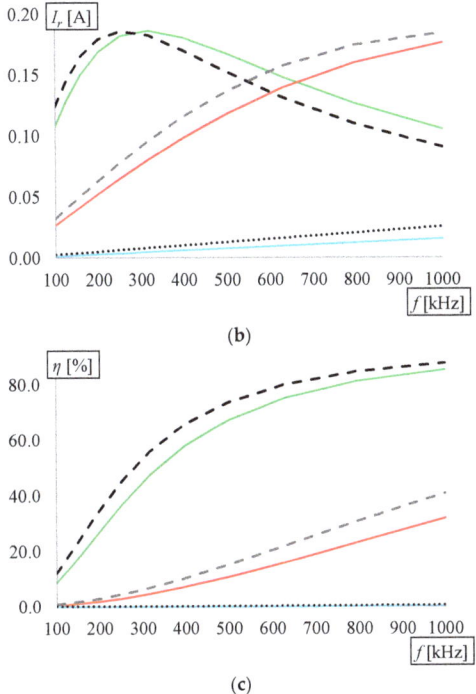

Figure 10. Results for the case $r_c = 20$ mm, $n_c = 30$: (**a**) transmitter current, (**b**) receiver current, and (**c**) power transfer efficiency.

The highest efficiency and one of the best accuracies were obtained for $n_c = 70$ (Figure 11c). The relative difference between currents for the least accurate case was 21.7% ($h = 20$ mm, $f = 1$ MHz). For the other distances, h results from FM and EC converged acceptably. Additionally, an analysis of I_r (Figures 10b and 11b) at identical efficiencies, η, showed that higher power was transferred to the load when coils with $n_c = 30$ were used. However, coils with $n_c = 70$ achieved $\eta \geq 80\%$ at lower frequencies.

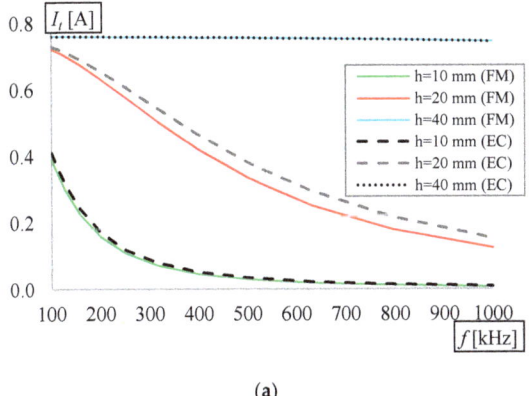

(**a**)

Figure 11. *Cont.*

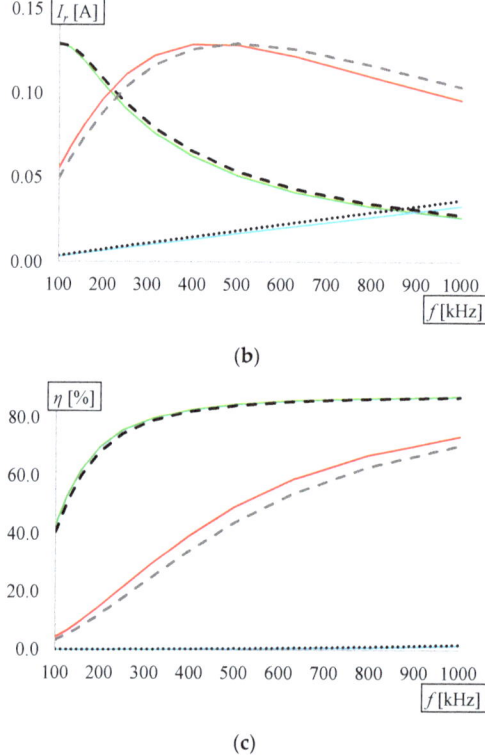

Figure 11. Results for the case $r_c = 20$ mm, $n_c = 70$: (**a**) transmitter current, (**b**) receiver current, and (**c**) power transfer efficiency.

The divergence between characteristics from the field and circuit models, and at the same time, the accuracy of the circuit analysis was expressed by root-mean-square deviation of the TR and RE currents.

$$\text{RMSD}_I = \sqrt{\frac{1}{2N_f}\left[\sum_{i=1}^{N_f}\left(\frac{I_{t,i}^{FM} - I_{t,i}^{EC}}{I_{t,i}^{FM}}\right)^2 + \sum_{i=1}^{N_f}\left(\frac{I_{r,i}^{FM} - I_{r,i}^{EC}}{I_{r,i}^{FM}}\right)^2\right]} \cdot 100\%, \qquad (23)$$

where $N_f = 10$ was the number of frequencies for which the calculations were made. RMSD_I was the combined difference between both currents at the entire considered bandwidth. The highest values (above 20%) were observed for $n_c = 30$ (small and large coil) at $h = 2r_c$ (Figure 12). Similar results were obtained for $r_c = 5$ mm, $n_c = 20$ (Figure 12a), and $r_c = 20$ mm, $n_c = 50$ (Figure 12b); however, RMSD_I was less than 20%. Discussed cases were related to systems, where the energy transfer efficiency was the order of a single percent (Figure 8c, Figure 9a–c, Figure 10a–c, Figure 11c); hence, presented differences had negligible practical significance. For the other cases, RMSD_I varied from 1.8% to 19.3%, and in eight variants, it was less than 10%. The circuit model provided a high degree of compliance, especially for $h/r_c < 2$, which were the distances between TR and RE coils, where the WPT system had the highest efficiency. Mean deviation for coil $r_c = 5$ mm was 11.5%, and for $r_c = 20$ mm, it was 13.3%. Obtained values indicated that the circuit model had comparable accuracy, despite the usage of smaller or larger coils. Thus, the proposed model can be used for an analysis of WPT cells with different sizes and numbers of turns.

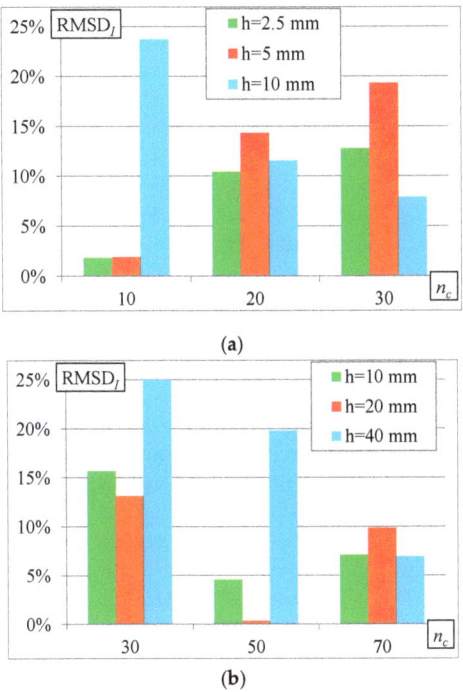

Figure 12. Root-mean-square deviation of the coils currents for considered frequency bandwidth and different WPT cell geometry: (**a**) $r_c = 5$ mm and (**b**) $r_c = 20$ mm.

3.3. Horizontal Misalignment

Additionally, an analysis of the horizontal misalignment (Δd) in the discussed periodic WPT system was performed. The numerical model was utilized to define the impact of Δd on relative change of power transfer efficiency η/η_{max}, where η_{max} is the transfer efficiency for $\Delta d = 0$. Two regions have been distinguished: area inside (A1) and outside (A2) the unit cell, as shown in Figure 13a. Computations at source frequency $f = 1$ MHz were performed for small-scale ($r_c = 5$ mm, $n_c = 20$) and large-scale coils ($r_c = 20$ mm, $n_c = 50$), where two distances ($h = 0.5 r_c$ and $h = r_c$) were considered.

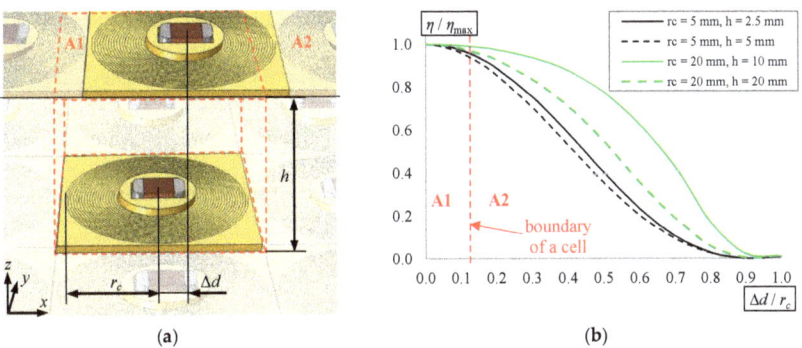

Figure 13. Horizontal misalignment in periodic WPT system: (**a**) visualization of horizontal distortion in WPT cell and (**b**) relative power transfer efficiency for different relative misalignment.

Horizontal misalignment has a relatively small impact on power transfer efficiency (Figure 13b), when the transmitter or receiver coil remains inside the WPT cell (area A1). Efficiency is slightly smaller (η/η_{max} = 0.981 at $\Delta d/r_c$ = 0.1) for large-scale coils at close distance ($h = 0.5r_c$); however, for smaller coils or at greater distance decreases faster (e.g., for $r_c = h = 5$ mm, η/η_{max} = 0.961 at $\Delta d/r_c$ = 0.1). Still, it may be assumed, that for a misalignment smaller than a boundary of a cell (in this case $\Delta d/r_c$ < 0.125), power transfer efficiency remains at a similar level, $\eta \approx \eta_{max}$. On the other hand, in an area A2, power transfer efficiency tends to be an almost-zero value, at $\Delta d/r_c$ = 1. The most "resistant" to misalignment, similarly as before, was the WPT cell with larger coils, especially at close distance. In this case, even a significant move of a coil beyond a cell's boundary ($\Delta d/r_c$ = 0.5) will reduce efficiency to η/η_{max} = 0.776, while for $r_c = h = 5$ mm, it will be more than two times smaller (η/η_{max} = 0.356).

4. Conclusions

The periodic wireless power transfer system was investigated by using numerical and circuit analysis. The authors defined the methodology of creating a field model of the WPT system, combined with current sheet approximation of multi-turn coils. The equivalent electrical circuit model of the WPT cell was proposed, which is an alternative for complex numerical analysis or experimental research of physical prototypes. The proposed circuit model provides the ability to perform fast and simplified calculations of WPT cells with different structures of coils. It is also possible to adjust electrical parameters of the system by utilizing the proposed models in order to design a periodic WPT structure with desired properties.

The introduced circuit model can replace the 3D field model, when analysis of periodic systems with many WPT cells is considered. The results indicated acceptable accordance of both models. Mean difference for computed variants of WPT system was 12.44%, with a standard deviation of 9.97%. This confirmed a possibility of estimating lumped parameters of the system by using the presented analytical formulas. A further analysis of WPT will focus on coils with various shapes and capacitive loads.

Author Contributions: The paper was written by A.S. The methodology and results presented in this paper were developed by A.S. and J.M.S. The analysis was performed by J.M.S. and A.C. The review, editing, and improvements to the content were made by A.C. All authors have read and agreed to the published version of the manuscript.

Funding: This work was supported by the Ministry of Science and Higher Education in Poland, at the Białystok University of Technology, under research subsidy No. WZ/WE-IA/2/2020.

Conflicts of Interest: The authors declare no conflict of interest.

References

1. Barman, S.D.; Reza, A.W.; Kumar, N.; Karim, M.E.; Munir, A.B. Wireless powering by magnetic resonant coupling: Recent trends in wireless power transfer system and its applications. *Renew. Sust. Energ. Rev.* **2015**, *51*, 1525–1552. [CrossRef]
2. Osowski, S.; Siwek, K. Data mining of electricity consumption in small power region. In Proceedings of the 19th International Conference Computational Problems of Electrical Engineering (CPEE), Banska Stiavnica, Slovakia, 9–12 September 2018; pp. 1–4.
3. Luo, Z.; Wei, X. Analysis of square and circular planar spiral coils in wireless power transfer system for electric vehicles. *IEEE Trans. Ind. Electron.* **2018**, *65*, 331–341. [CrossRef]
4. Batra, T.; Schaltz, E.; Ahn, S. Effect of ferrite addition above the base ferrite on the coupling factor of wireless power transfer for vehicle applications. *J. Appl. Phys.* **2015**, *117*. [CrossRef]
5. Eteng, A.A.; Rahim, S.K.A.; Leow, C.Y.; Chew, B.W.; Vandenbosch, G.A.E. Two-stage design method for enhanced inductive energy transmission with Q-constrained planar square loops. *PLoS ONE* **2016**, *11*, e0148808. [CrossRef]
6. Kim, T.-H.; Yun, G.-H.; Lee, W.Y. Asymmetric coil structures for highly efficient wireless power transfer systems. *IEEE Trans. Microw. Theory. Tech.* **2018**, *66*, 3443–3451. [CrossRef]
7. Rim, C.T.; Mi, C. *Wireless Power Transfer for Electric Vehicles and Mobile Devices*; John Wiley & Sons, Ltd.: Hoboken, NJ, USA, 2017; pp. 473–490.

8. Fujimoto, K.; Itoh, K. *Antennas for Small Mobile Terminals*, 2nd ed.; Artech House: Norwood, MA, USA, 2018; pp. 30–70.
9. Zhang, Z.; Pang, H.; Georgiadis, A.; Cecati, C. Wireless power transfer—An overview. *IEEE Trans. Ind. Electron.* **2019**, *66*, 1044–1058. [CrossRef]
10. Rozman, M.; Fernando, M.; Adebisi, B.; Rabie, K.M.; Collins, T.; Kharel, R.; Ikpehai, A. A new technique for reducing size of a WPT system using two-loop strongly-resonant inductors. *Energies* **2017**, *10*, 1614. [CrossRef]
11. Liu, X.; Wang, G. A novel wireless power transfer system with double intermediate resonant coils. *IEEE Trans. Ind. Electron.* **2016**, *63*, 2174–2180. [CrossRef]
12. El Rayes, M.M.; Nagib, G.; Abdelaal, W.G.A. A review on wireless power transfer. *IJETT* **2016**, *40*, 272–280. [CrossRef]
13. Re, P.D.H.; Podilchak, S.K.; Rotenberg, S.; Goussetis, G.; Lee, J. Circularly polarized retrodirective antenna array for wireless power transmission. In Proceedings of the 11th European Conference on Antennas and Propagation (EUCAP), Paris, France, 19–24 March 2017; pp. 891–895.
14. Nikoletseas, S.; Yang, Y.; Georgiadis, A. *Wireless Power Transfer Algorithms, Technologies and Applications in Ad Hoc Communication Networks*; Springer: Cham, Switzerland, 2016; pp. 31–51.
15. Stevens, C.J. Magnetoinductive waves and wireless power transfer. *IEEE Trans. Power Electron.* **2015**, *30*, 6182–6190. [CrossRef]
16. Zhong, W.; Lee, C.K.; Hui, S.Y.R. General analysis on the use of Tesla's resonators in domino forms for wireless power transfer. *IEEE Trans. Ind. Electron.* **2013**, *60*, 261–270. [CrossRef]
17. Alberto, J.; Reggiani, U.; Sandrolini, L.; Albuquerque, H. Accurate calculation of the power transfer and efficiency in resonator arrays for inductive power transfer. *PIER* **2019**, *83*, 61–76. [CrossRef]
18. Alberto, J.; Reggiani, U.; Sandrolini, L.; Albuquerque, H. Fast calculation and analysis of the equivalent impedance of a wireless power transfer system using an array of magnetically coupled resonators. *PIER B* **2018**, *80*, 101–112. [CrossRef]
19. Martin, P.; Ho, B.J.; Grupen, N.; Muñoz, S.; Srivastasa, M. An iBeacon primer for indoor localization. In Proceedings of the 1st ACM Conference on Embedded Systems for Energy-Efficient Buildings (BuildSys'14), Memphis, TN, USA, 3–6 November 2014; pp. 190–191.
20. Li, X.; Zhang, H.; Peng, F.; Li, Y.; Yang, T.; Wang, B.; Fang, D. A wireless magnetic resonance energy transfer system for micro implantable medical sensors. *Sensors* **2012**, *12*, 10292–10308. [CrossRef] [PubMed]
21. Kim, D.; Abu-Siada, A.; Sutinjo, A. State-of-the-art literature review of WPT: Current limitations and solutions on IPT. *Electr. Power Syst. Res.* **2018**, *154*, 493–502. [CrossRef]
22. Fitzpatrick, D.C. *Implantable Electronic Medical Devices*; Academic Press: San Diego, CA, USA, 2014; pp. 7–35.
23. Wang, B.; Yerazunis, W.; Teo, K.H. Wireless power transfer: Metamaterials and array of coupled resonators. *Proc. IEEE* **2013**, *101*, 1359–1368. [CrossRef]
24. Kanoun, O. *Lecture Notes on Impedance Spectroscopy*; CRC Press: Boca Raton, FL, USA, 2015; Volume 5, pp. 63–71.
25. Meeker, D.C. Continuum Representation of Wound Coils Via an Equivalent Foil Approach. Available online: http://www.femm.info/examples/prox/notes.pdf (accessed on 3 April 2020).
26. Meeker, D.C. An improved continuum skin and proximity effect model for hexagonally packed wires. *J. Comput. Appl. Math.* **2012**, *236*, 4635–4644. [CrossRef]
27. Mohan, S.S.; del Mar Hershenson, M.; Boyd, S.P.; Lee, T.H. Simple Accurate expressions for planar spiral inductances. *IEEE J. Solid-State Circuits* **1999**, *34*, 1419–1424. [CrossRef]
28. Raju, S.; Wu, R.; Chan, M.; Yue, C.P. Modeling of mutual coupling between planar inductors in wireless power applications. *IEEE Trans. Power Electron.* **2014**, *29*, 481–490. [CrossRef]
29. Tal, N.; Morag, Y.; Levron, Y. Magnetic induction antenna arrays for MIMO and multiple-frequency communication systems. *PIER C* **2017**, *75*, 155–167. [CrossRef]
30. Liu, S.; Su, J.; Lai, J. Accurate expressions of mutual inductance and their calculation of archimedean spiral coils. *Energies* **2019**, *12*, 2017. [CrossRef]

 © 2020 by the authors. Licensee MDPI, Basel, Switzerland. This article is an open access article distributed under the terms and conditions of the Creative Commons Attribution (CC BY) license (http://creativecommons.org/licenses/by/4.0/).

Article

Parameter Analysis and Optimization of Class-E Power Amplifier Used in Wireless Power Transfer System

Feng Wen [1,2,*] and Rui Li [1]

1. School of Automation, Nanjing University of Science and Technology, Nanjing 210094, China
2. Jiangsu Provincial Key Laboratory of Smart Grid Technology and Equipment, Nanjing 210096, China
* Correspondence: wen@njust.edu.cn; Tel.: +86-159-9620-0950

Received: 18 July 2019; Accepted: 20 August 2019; Published: 22 August 2019

Abstract: In this paper, a steady-state matrix analysis method is introduced to analyze the output characteristics of the class-E power amplifier used in a wireless power transfer (WPT) system, which takes the inductance resistance, on-resistance and leakage current of metal-oxide-semiconductor field effect transistor (MOSFET) into account so that the results can be closer to the actual value. On this basis, the parameters of the class-E power amplifier are optimized, and the output power is improved under the premise of keeping the efficiency unchanged. Finally, the output characteristics of the amplifier before and after optimization are compared by an experiment, while the B-field strength around the WPT system is studied through simulation. The experimental results verify the correctness and feasibility of the optimization method based on steady-state matrix analysis.

Keywords: steady-state matrix analysis; Class-E power amplifier; wireless power transfer (WPT) system; output characteristics; strength

1. Introduction

The Class-E power amplifier is widely used in high-frequency power supply, wireless power transfer (WPT) and other fields because of its simple structure and high output efficiency [1,2]. In 2007, the research team of Massachusetts Institute of Technology (MIT) put forward the magnetic coupling resonant wireless power transfer technology, and the class-E power amplifier has once again become a hot research topic at home and abroad [3].

MOSFET can meet the zero-voltage switching (ZVS) and zero-derivative switching (ZDS) conditions when the class-E power amplifier works under the ideal condition and the efficiency is 100% [4]. According to the parametric characters of the class-E power amplifier, the authors studied the changes in output characteristics of class-E power amplifiers when the load deviated from the optimal value in [5]. The relationship between the DC power supply of a class-E power amplifier and MOSFET's peaks voltage is studied in [6]. The effect of various parameters of a class-E power amplifier on the output characteristics of the circuit, the voltage and the current waveform of the MOSFET are analyzed in [7]. T. Mury and his team conducted an in-depth study on the operating characteristics of class-E power amplifiers in the sub-optimal working state where the duty cycle of the MOSFET is not 50%, the mathematical modeling of the class-E power amplifier is carried out, and the influence of duty cycle on the current peak, output voltage and current is analyzed [8]. The analysis of a class-E power amplifier based on a lossless switch and ratio-frequency (RF) choke (RFC) was introduced in [9]. In [10] and [11], researchers used resonant soft-switching converters to achieve optimum switching conditions. Class-EF inverters and the equations of the voltages and currents were derived with traditional analysis in [12]. In [13], a novel topology of the Class-E_M power amplifier was proposed based on a finite direct current (DC)-feed inductance and an isolation circuit. R. A. Beltran et al., proposed a simplified analysis and design of class-E outphasing transmitters that predicts efficiency and output power as a function of input drive phase difference based on classic design equations of the class-E amplifier [14]. However, the analytical processes of the traditional class-E power amplifier circuit above are ideal, generally ignoring the MOSFET's turn-on resistance and considering the turn-off resistance to be infinite. In other words, analysis of every component in load network is based on the complete ideal condition such as no impedance [15]. There is a certain difference between the theoretical result and the actual value where the performance of the class-E power amplifier can be significantly influenced by the non-ideal factor such as the inductance resistance, finite dc-feed inductance, leakage current and so on [16].

In this paper, a steady-state matrix analysis method which proved to be a fast and effective approach for optimization of switching-mode power amplifiers [17–20] is used for studying the output characteristics of two kinds of class-E power amplifiers (a traditional choke, RFC type [21], parallel load type, parallel capacitor and inductor (PCL) type [22]) under the consideration of non-ideal factors, such as the inductance resistance, leakage current and so on. On the basis of inductance of phase angle compensation L_x (L_{in}), the influence of the parallel resonant capacity C_p and the load R_L on the output power, the operation efficiency and the maximum withstand voltage of MOSFET is turned off, two kinds of class-E power amplifiers are optimized. The output power of the class-E power amplifier is improved on ensuring output efficiency and the effectiveness of optimization of class-E power amplifier circuit based on steady-state matrix analysis is verified according to the experiments. The B-field strength around the WPT system is also simulated and studied.

2. Application of Steady-State Matrix Analysis Method

The equivalent circuit of the traditional class-E power amplifier is shown in Figure 1.

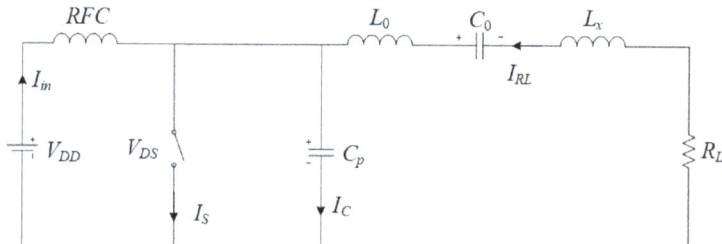

Figure 1. Class-E model under ideal condition.

The average value of the voltage on the MOSFET during one period is equal to the DC power supply, V_{DD}:

$$V_{DD} = \frac{1}{2\pi} \int_0^{2\pi} V_{DS}(\omega t) d\omega t = \frac{\pi I_{in}}{\omega C_p} \quad (1)$$

In the ideal case, the efficiency of the Class-E power amplifier is 100%, where the active power absorbed by the load is equal to the input provided by the DC power supply:

$$I_{in} V_{DD} = \frac{I_{RL}^2}{2} R_L \quad (2)$$

$$P_o = 0.5768 \frac{V_{DD}^2}{R_L} \quad (3)$$

The voltage applied to R_L and L_x is the fundamental frequency voltage, which can be obtained by Fourier analysis:

$$\begin{cases} V_R = -\frac{1}{\pi}\int_0^{2\pi} V_{DS}(\omega t)\sin(\omega t)d\omega t \\ V_{Lx} = -\frac{1}{\pi}\int_0^{2\pi} V_{DS}(\omega t)\cos(\omega t)d\omega t \end{cases} \quad (4)$$

The parameter calculation formula of the class-E power amplifier load network can be derived from (4):

$$\frac{\omega L_x}{R_L} = 1.1525 \quad (5)$$

$$\omega C_p R_L = 0.1836 \quad (6)$$

The maximum withstand voltage on the MOSFET is:

$$V_{ds\_max} = 3.562 V_{DD} \quad (7)$$

MOSFET's turn-on resistance, leakage current and inductance resistance in the load network are considered in the circuit diagram of a class-E power amplifier in Figure 2. MOSFET's loss can be divided into the on-resistance loss and the leakage current loss in turn-off state. Inductance resistance comes from choke (parallel inductor) L_{in} resonant inductor L_0 and inductance of phase angle compensation L_x.

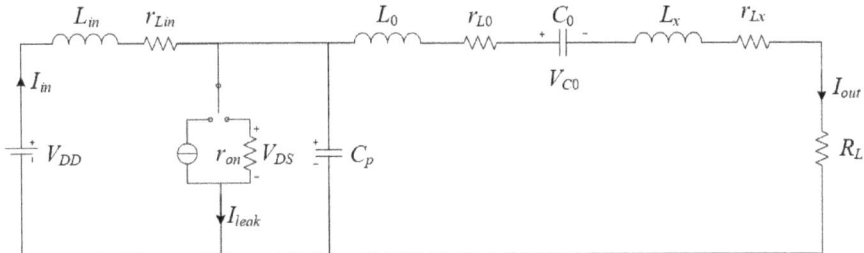

Figure 2. Class-E model under non-ideal condition.

The state variables in the diagram is defined as the following matrix:

$$q(t) = \begin{bmatrix} I_{in}(t) & I_{out}(t) & V_{DS}(t) & V_{CO}(t) \end{bmatrix}^T, \quad (8)$$

$I_{in}(t)$ is the input current of DC power supply, $I_{out}(t)$ is the output current, $V_{DS}(t)$ is drain-source voltage, $V_{CO}(t)$ is the voltage of the resonant capacitor. They are energy storage variables in the circuit, Therefore, the switching state of the MOSFET will not cause changes immediately, which meets the requirement of steady-state analysis.

The state equation of class-E power amplifier circuit can be obtained by the first derivative of the state variable. When MOSFET is turned off, the equation of state is:

$$\begin{cases} L_{in}\frac{dI_{in}(t)}{dt} = V_{DD} - V_{DS}(t) - r_{Lin}I_{in}(t) \\ (L_0 + L_x)\frac{dI_{out}(t)}{dt} = V_{DS}(t) - (r_{L0} + r_{Lx} + R_L)I_{out}(t) - V_{CO}(t) \\ C_p\frac{dV_{DS}(t)}{dt} = I_{in}(t) - I_{out}(t) - I_{leak} \\ C_O\frac{dV_{CO}(t)}{dt} = I_{out}(t) \end{cases} \quad (9)$$

Changing I_{leak} to V_{DS}/r_{on} in (9), the equation of state when MOSFET is turned on is obtained. All the equations of state have the form of first-order differential equations:

$$\frac{dq(t)}{dt} = Aq(t) + B. \quad (10)$$

The state of matrix A and matrix B have different forms when the MOSFET is turned off and turned on. For the convenience of analysis, the state matrix when MOSFET is turned off is defined as A_1 and B_1, the off-time is t_1. Moreover, the state matrix when MOSFET is turned on is defined as A_2 and B_2, and the on-time is t_2. The solution of (10) is:

$$q(t) = e^{At}q_0 + \int_0^t e^{A(t-\tau)}B d\tau = e^{At}q_0 + A^{-1}\left(e^{At}-I\right)B, \qquad (11)$$

q_0 is the initial value. The solution of the state equation is:

$$q_1(t) = e^{A_1 t}q_{01} + A_1^{-1}\left(e^{A_1 t}-I\right)B_1, \qquad (12)$$

$$q_2(t) = e^{A_2 t}q_{02} + A_2^{-1}\left(e^{A_2 t}-I\right)B_2. \qquad (13)$$

Because of the energy properties of the state variable, there is no immediate change of state variable when MOSFET works from turn-off to turn-on. As the MOSFET is turned off, the state variable at t_1 time is equal to the value at $t = 0$ that is q_{01}. The following equations can be derived:

$$q_{02} = q_1(t_1) = e^{A_1 t_1}q_{01} + A_1^{-1}\left(e^{A_1 t_1}-I\right)B_1, \qquad (14)$$

$$q_{01} = q_2(t_2) = e^{A_2 t_2}q_{02} + A_2^{-1}\left(e^{A_2 t_2}-I\right)B_2. \qquad (15)$$

The initial value q_{01}, q_{02} can be obtained in (14) and (15), losses and output power can be defined:

$$W_1 = \int_0^{t_1} q_1 q_1^T dt, \qquad (16)$$

$$W_2 = \int_0^{t_2} q_2 q_2^T dt. \qquad (17)$$

Output power:

$$P_{OUT} = \frac{R_L}{T}\int_0^T I_{OUT}^2(t)dt = \frac{R_L}{T}\{W_1[2,2] + W_2[2,2]\}. \qquad (18)$$

Loss of the inductor resistance:

$$P_{rL0} = \frac{r_{L0}}{T}\{W_1[2,2] + W_2[2,2]\}, \qquad (19)$$

$$P_{rLX} = \frac{r_{LX}}{T}\{W_1[2,2] + W_2[2,2]\}, \qquad (20)$$

$$P_{rLin} = \frac{r_{Lin}}{T}\{W_1[1,1] + W_2[1,1]\}. \qquad (21)$$

Loss of MOSFET in turn-on and turn-off state are shown as follows:

$$P_{Son} = \frac{1}{T}C_{on}\left(\int_0^{t_2} q_2 q_2^T dt\right)C_{on}^T, \qquad (22)$$

$$P_{Soff} = \frac{1}{T}C_{off}\left(\int_0^{t_1} q_1 dt\right)C_{off}^T, \qquad (23)$$

$$C_{on} = \begin{bmatrix} 0 & 0 & 1/\sqrt{R_{on}} & 0 \end{bmatrix}, \qquad (24)$$

$$C_{off} = \begin{bmatrix} 0 & 0 & \sqrt{I_{leak}} & 0 \end{bmatrix}. \qquad (25)$$

The output power of class-E power amplifier, efficiency and drain-source voltage of MOSFET can be obtained according to (16)–(25).

3. Analysis of the Characteristic of Class-E Power Amplifier

Based on the steady-state matrix analysis method above, the effect of the inductor phase angle compensation L_x (L_{in}) and parallel capacitance C_p on the output characteristics of two kinds of class-E power amplifier can be studied. The theoretical parameter values of two types with a DC power supply of 25 V, working at 6 MHz and output power of 30 W are shown in Table 1. $R_{L\text{-}og}$, L_{X_og}, L_{0_og}, and C_{0_og} are theoretical values of load resistance, inductance of phase angle compensation, inductance and capacitance of resonant circuit. L_{in_og} and C_{p_og} are theoretical values of choke (parallel) inductance and capacitance. r_{LX}, r_{Lin}, and r_{L0} are resistance values of the inductors.

Table 1. Theoretical parameter values of class-E.

Parameter	RFC Type	PCL Type
$R_{L\_og}(\Omega)$	12	28.4
$L_{X\_og}(\mu H)$	0.367	0
$r_{LX}(\Omega)$	0.19	0
$C_{p\_og}(nF)$	0.405	0.639
$L_{in\_og}(\mu H)$	15	0.552
$r_{Lin}(\Omega)$	0.27	0.31
$C_{0\_og}(nF)$	0.368	0.155
$L_{0\_og}(\mu H)$	1.91	4.53
$r_{L0}(\Omega)$	0.45	0.87

In order to analyze the effect of L_x (L_{in}) and C_p on the output characteristics of class-E power amplifier, the surface graph of its parametric characters is obtained based on the steady-state matrix method. In the following analysis, MRF6V2150NBR1 is used so that MOSFET's turn-on resistance R_{on} is 0.3 Ω, the leakage current I_{leak} is 2.5 mA, and drain source maximum withstand voltage is 110 V.

3.1. Influence of RFC Output Characteristics

The effect of phase angle compensation inductance L_x which is parallel with capacitance C_p and the load R_L on output power, efficiency and MOSFET's peak voltage in RFC type is shown in Figures 3 and 4. The variation range of L_x is from 0.27 µH to 0.47 µH, C_p is from 300 pF to 500 pF and R_L is from 8 Ω to 16 Ω. If MOSFET's peaks voltage exceeds the maximum withstand voltage, it will cause damage to MOSFET. Therefore, the maximum withstand voltage should also be considered in designing a class-E power amplifier.

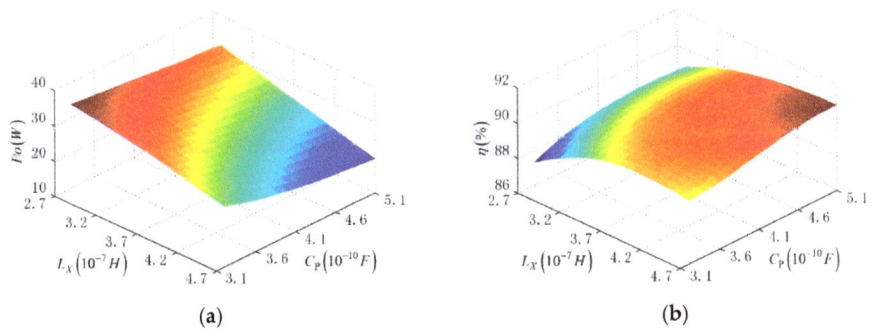

Figure 3. Cont.

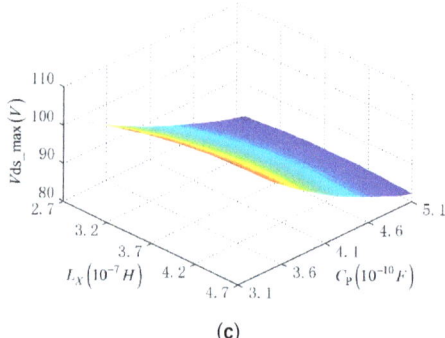

(c)

Figure 3. (a) Output power of RFC changing with L_x and C_p; (b) output efficiency of RFC changing with L_x and C_p; (c) peak voltage of RFC changing with L_x and C_p.

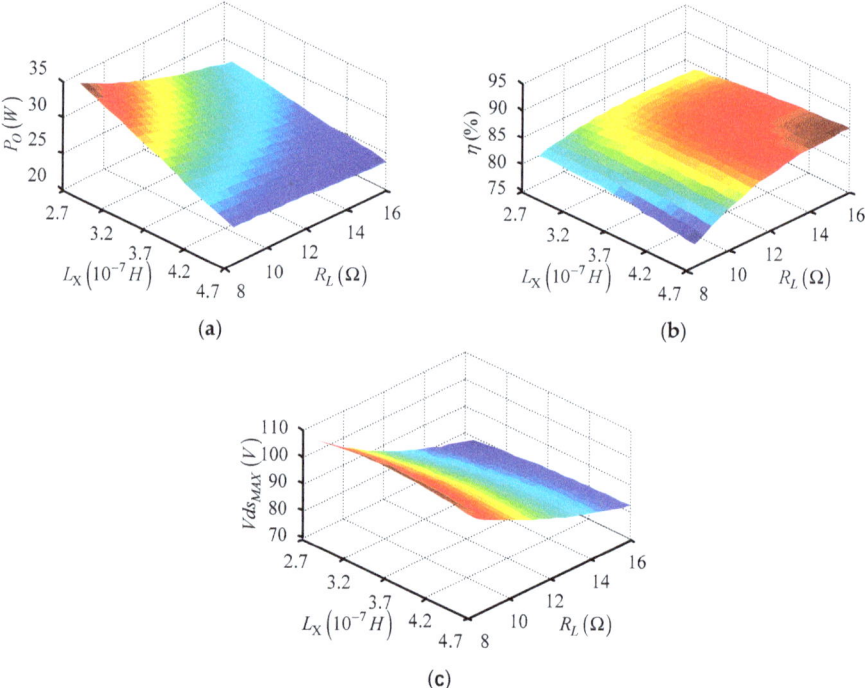

Figure 4. (a) Output power of RFC changing with L_x and R_L; (b) output efficiency of RFC changing with L_x and R_L; (c) peak voltage of RFC changing with L_x and R_L.

With the increase of L_x and C_p, the output power of the RFC type decreases significantly, the efficiency increases firstly and then decreases. The trend of output efficiency is opposite to the trend of output power, and the influence of the two factors should be considered comprehensively in the design of circuit parameters. Moreover, L_x has little effect on the maximum voltage of MOSFET, and its value decreases mainly with the increase of C_p, which can be obtained from Figure 3.

It can be seen from Figure 4 that the output power of the class-E power amplifier decreases while the efficiency increases with the increase of R_L. The effect of load R_L on output power is obvious, while the trend of the output efficiency is opposite to that of the output power. On the other hand, L_x has less influence on the maximum voltage of MOSFET than R_L. However, within the variation range, the voltage peak across the MOSFET is smaller than its maximum value, ensuring that the MOSFET can work normally without being broken down.

3.2. Influence of PCL Characteristics

The effect of phase angle compensation inductance L_{in} on output power, efficiency and MOSFET peaks voltage in PCL type is shown in Figure 5. The variation range of L_{in} is from 0.45 µH to 0.65 µH, C_p is from 550 pF to 750 pF.

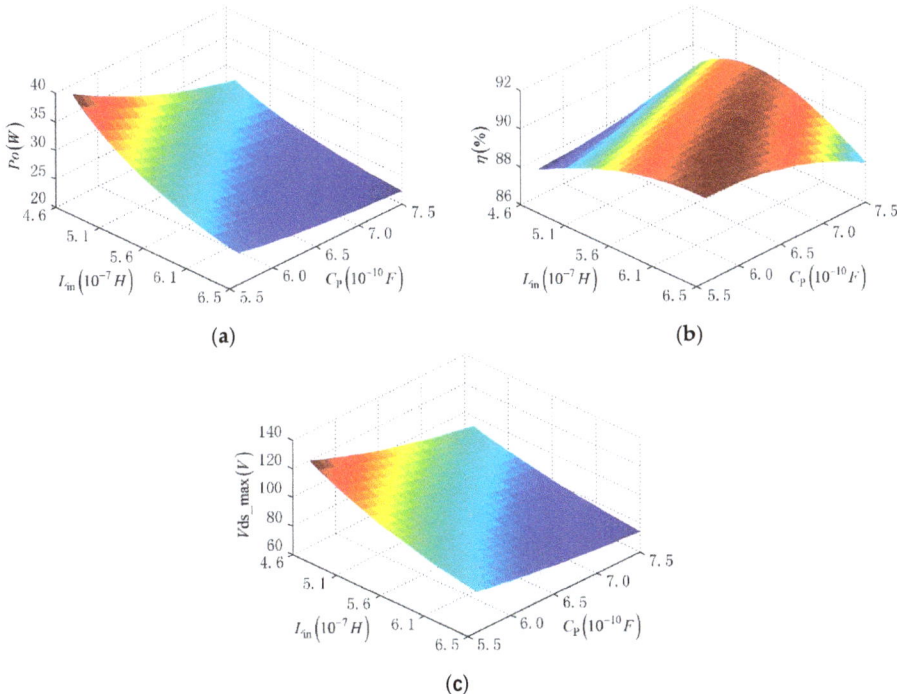

Figure 5. (a) Output power of PCL changing with L_{in} and C_p; (b) output efficiency of PCL changing with L_{in} and C_p; (c) peak voltage of PCL changing with L_{in} and C_p.

With the increase of L_{in} and C_p, the output power of the PCL type also decreases as shown in Figure 5a; when the C_p value is fixed, the efficiency increases firstly and then decreases with the increase of L_{in}, when the L_{in} value is small, the efficiency increases with the increase of C_p, while the L_{in} value becomes larger, the efficiency gradually turns to decrease with the increase of C_p as shown in Figure 5b; the maximum withstand voltage of PCL type decreases significantly with the increase of L_{in} and C_p as shown in Figure 5c. When the L_{in} and C_p values are too small, the maximum withstand voltage will exceed 120 V, which will easily cause damage to the MOSFET.

3.3. Optimization Strategy of Two Kinds of Class-E Power Amplifier

According to the analysis above, due to the existence of inductance resistance and non-ideal MOSFET, the designed circuit parameters are no longer optimal under ideal condition. The model of class-E power amplifiers using state equations is closer to the real situation. For the design of a broadband class-E power amplifier, if the reactance corresponding to the phase shifting inductor L_x is regarded as part of the load impedance, the circuit can be optimized overall by adjusting the phase shifting inductor L_x (L_{in}), C_p and R_L under the condition of working at 6 MHz and output power of 30 W.

The RFC type parameters before and after optimization are shown in Table 2. The ideal values are calculated according to (1)–(7). As we select L_x = 0.34 μH, C_p = 0.38 nF, R_L = 9.12 Ω according to Figures 3 and 4, the output efficiency of the optimized class-E power amplifier has no obvious change, the output power is increased by 9.25% (2.6 W). Moreover, the maximum withstand voltage of MOSFET is slightly increased in turn-off stage.

Table 2. Parameter values of RFC before and after optimization.

Parameter	Ideal Value	Optimized Value
$L_x(\mu H)$	0.367	0.34
$C_p(nF)$	0.405	0.38
$R_L(\Omega)$	12	9.12
$P_o(W)$	28.11	30.71
$\eta(\%)$	90.32	90.03
$Vds_{max}(V)$	92.41	95.01

The PCL type parameters before and after optimization are shown in Table 3. Select L_{in} = 0.54 μH, C_p = 0.6 nF. Under the condition that the output efficiency of the optimized class-E power amplifier has no obvious change, the output power is increased by 7.61% (2.02 W). Moreover, the maximum withstand voltage of MOSFET in turn-off stage is increased significantly, still within the safe range.

Table 3. Parameter values of PCL before and after optimization.

Parameter	Ideal Value	Optimized Value
$L_{in}(\mu H)$	0.552	0.54
$C_p(nF)$	0.639	0.6
$P_o(W)$	26.55	28.57
$\eta(\%)$	90.33	89.94
$Vds_{max}(V)$	90.35	96.90

4. Experimental Results and Analysis

Comparing the output character of two kinds of class-E power amplifier before and after optimization, the validity of the optimization method is verified using the parameters obtained from the analysis. The experimental platform is built as shown in Figure 6.

Figure 6. Experimental platform.

4.1. Experiments of RFC Type under Non-Ideal Condition

The experimental waveforms of RFC type before and after optimization is shown in Figure 7. Before optimization, the RMS of the output current, the output voltage and the output power of RFC type were 1.48 A, 18.8 V and 27.82 W. The operational efficiency was 85.24%. After optimization, the RMS of the output current, the output voltage and the output power became 1.56 A, 19.3 V and 30.11 W. The operational efficiency was 84.76%.

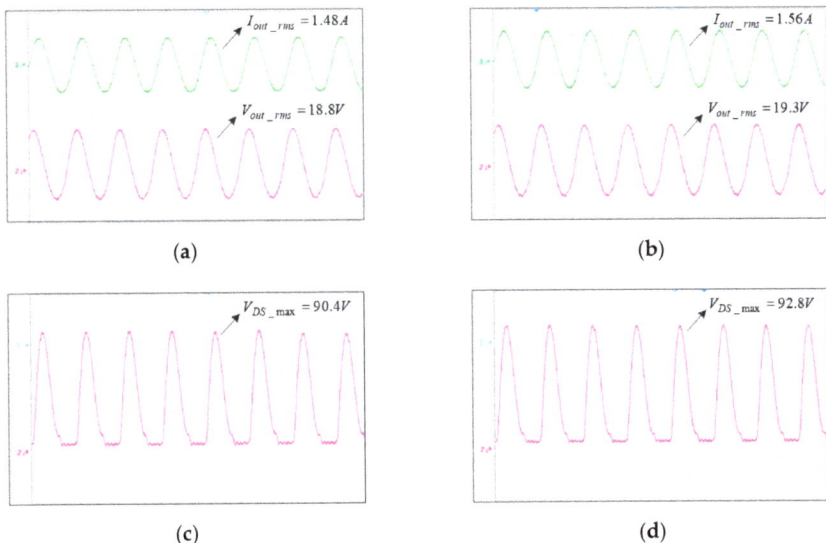

Figure 7. (a) Load current and voltage waveform of RFC before optimization; (b) load current and voltage waveform of RFC after optimization; (c) drain-source voltage waveform of PCL before optimization; (d) drain-source voltage waveform of PCL after optimization.

Under the condition that the efficiency has no obvious change, the output power is increased by 2.29 W (8.23%).

The maximum withstand voltage of MOSFET in turn-off stage increases from 90.4 V to 92.8 V, which is still within the safe range. Moreover, due to the small voltage oscillation in turn-on stage which increases the loss of the MOSFET, the experimental output efficiency is smaller than the theoretical value.

4.2. Experiments of PCL Type under Non-Ideal Condition

The experimental waveforms of PCL type before and after optimization are shown in Figure 8. Before optimization, the RMS of the output current of PCL type was 0.97 A, the output voltage was 26.8 V and the output power was 25.99 W; the efficiency was 85.57%. After optimization, the RMS of the output current of RFC type was 1.0 A, the output voltage was 27.9 V, and the output power was 27.90 W; the efficiency was 85.84%. Under the condition that the output efficiency had no obvious change, the output power increased by 1.91 W (7.35%). The maximum withstand voltage of MOSFET was increased from 88.8 V to 95.2 V, but still within the safe range.

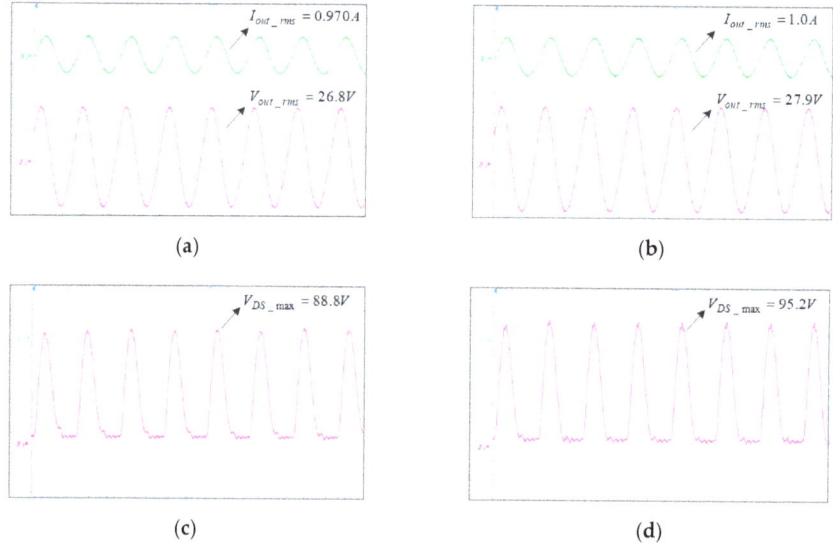

Figure 8. (**a**) Load current and voltage waveform of PCL before optimization; (**b**) load current and voltage waveform of PCL after optimization; (**c**) drain-source voltage waveform of PCL before optimization; (**d**) drain-source voltage waveform of PCL after optimization.

5. Simulation Result of B-Field Strength

A simulation model of wireless power transfer system was established according to the experimental platform as shown in Figure 9.

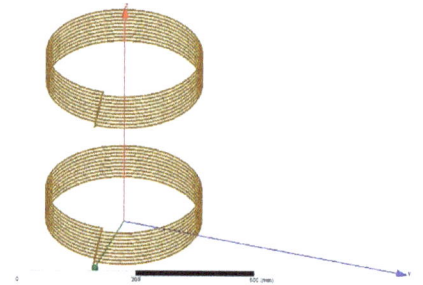

Figure 9. Simulation model of wireless power transfer system.

Four cases are considered in this paper: wireless power transfer system with RFC-type before optimization (Case 1); wireless power transfer system with RFC-type after optimization (Case 2); wireless power transfer system with PCL-type before optimization (Case 3); wireless power transfer system with PCL-type after optimization (Case 4). B-field strength results of the four cases are shown in Figure 10.

For further research on the B-field strength around the wireless power transfer system, we studied the B-field strength along the y-axis and the z-axis both from the specified point (0.0.3 m, 0.25 m) which was 0.1 m to the edge of the coils using the coordinate system in Figure 9.

The trends of the B-field strength along two lines in four cases are shown in Figure 11. 'I' represents B-field strength in case 1 along the y-axis; 'II' represents B-field strength in case 1 along the z-axis; 'III' represents B-field strength in case 2 along the y-axis; 'IV' represents B-field strength in case 2 along the z-axis; 'V' represents B-field strength in case 3 along the y-axis; 'VI' represents B-field strength in case 3 along the z-axis; 'VII' represents B-field strength in case 4 along the y-axis; 'VIII' represents B-field strength in case 4 along the z-axis. B-field strength around the system with the RFC-type is stronger than that with the PCL-type, while the B-field strength after optimization was stronger than that before optimization. In general, the B-field strength in the four cases was always less than 27 µT which is the reference level for general public exposure formulated by ICNIRP-2010.

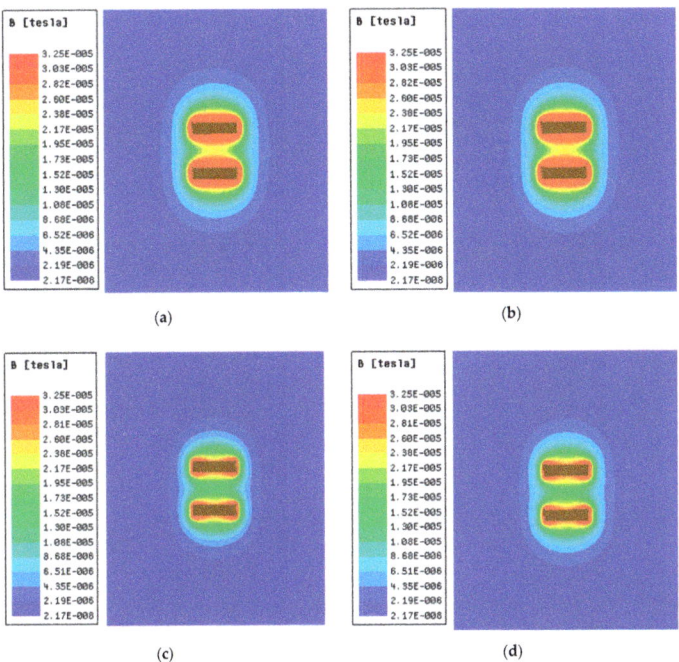

Figure 10. (**a**) Simulation result of B-field strength in case 1; (**b**) simulation result of B-field strength in case 2; (**c**) simulation result of B-field strength in case 3. (**d**) simulation result of B-field strength in case 4.

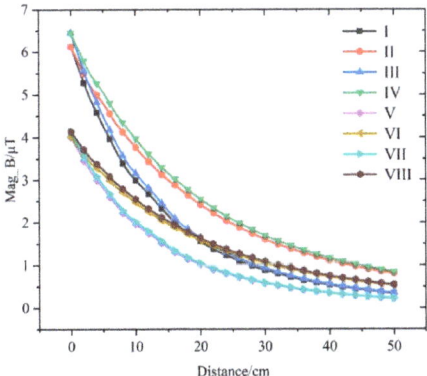

Figure 11. The trends of the B-field strength along two lines in four cases.

6. Conclusions

In this paper, a steady-state matrix analysis method suitable for class-E power amplifier was introduced. Compared with the analysis process of the working principle of a class-E power amplifier, non-ideal factors are considered in this method, such as inductance resistance and leakage current, etc. The experimental results are more accurate. Based on steady-state matrix analysis, the output characteristic of two kinds of class-E power amplifier circuits were analyzed and optimized. The experimental results show that the output power of the two types of class-E power amplifier increased by 2.29 W (8.23%) and 1.91 W (7.35%), respectively, while the output efficiency had no obvious change. The B-field strength of the systems with two types of class-E power amplifier before and after optimization meets the ICNIRP-2010 standard.

Author Contributions: F.W. conceived and designed the study, and this work was performed under the advice of and regular feedback from him. R.L. was responsible for the simulations.

Funding: This work was supported by the Basic Research Program of Jiangsu Province (No. BK20180485), Jiangsu Provincial Key Laboratory of Smart Grid Technology and Equipment Project, and the Fundamental Research Funds for the Central Universities (No. 30919011241).

Conflicts of Interest: The authors declare no conflict of interest.

References

1. Kkelis, G.; Yates, D.C.; Mitcheson, P.D. Class-E Half-Wave Zero dv/dt Rectifiers for Inductive Power Transfer. *IEEE Trans. Power Electron.* **2017**, *32*, 8322–8337. [CrossRef]
2. Chokkalingam, B.; Padmanaban, S.; Leonowicz, Z. Class E Power Amplifier Design and Optimization for the Capacitive Coupled Wireless Power Transfer System in Biomedical Implants. *Energies* **2017**, *10*, 1409.
3. Kwan, C.H.; Kkelis, G.; Aldhaher, S.; Lawson, J.; Yates, D.C.; Luk, P.C.K.; Mitcheson, P.D. Link efficiency-led design of mid-range inductive power transfer systems. In Proceedings of the 2015 IEEE PELS Workshop on Emerging Technologies: Wireless Power (2015 WoW), Daejeon Gwangyeoksi, Korea, 5–6 June 2015; pp. 1–7.
4. Cai, W.; Zhang, Z.; Ren, X.; Liu, Y.-F. A 30-MHz isolated push-pull VHFresonantconverter. In Proceedings of the 2014 IEEE Applied Power Electronics Conference and Exposition-APEC, Fort Worth, TX, USA, 16–20 March 2014; pp. 1456–1460.
5. Suetsugu, T.; Kazimierczuk, M. Analysis of transient behavior of class E amplifier due to load variations. In Proceedings of the 2011 IEEE Ninth International Conference on Power Electronics and Drive Systems, Singapore, 5–8 December 2011; pp. 600–603.
6. Jaimes, A.F.; de Sousa, F.R. Simple expression for estimating the switch peak voltage on the class-E amplifier with finite DC-feed inductance. In Proceedings of the 2016 IEEE 7th Latin American Symposium on Circuits & Systems (LASCAS), Florianopolis, Brazil, 28 February–2 March 2016; pp. 183–186.
7. Aldhaher, S.; Mitcheson, P.D.; Yates, D.C. Load-independent Class EF inverters for inductive wireless power transfer. In Proceedings of the 2016 IEEE Wireless Power Transfer Conference (WPTC), Aveiro, Portugal, 5–6 May 2016; pp. 1–4.
8. Mury, T.; Fusco, V. Exploring figures of merit associated with the suboptimum operation of class-E power amplifiers. *IET Circuits Devices Syst.* **2007**, *1*, 401–407. [CrossRef]
9. Raab, F. Idealized operation of the Class E tuned power amplifier. *IEEE Trans. Circuits Syst.* **1977**, *24*, 725–735. [CrossRef]
10. Aldhaher, S.; Kkelis, G.; Yates, D.C.; Mitcheson, P.D. Class EF2 inverters for wireless power transfer applications. In Proceedings of the 2015 IEEE Wireless Power Transfer Conference (WPTC), Boulder, CO, USA, 13–15 May 2015; pp. 1–4.
11. Choi, J.; Tsukiyama, D.; Tsuruda, Y.; Rivas, J. 13.56 MHz 1.3 kW resonant converter with GaN FET for wireless power transfer. In Proceedings of the 2015 IEEE Wireless Power Transfer Conference (WPTC), Boulder, CO, USA, 13–15 May 2015; pp. 1–4.
12. Aldhaher, S.; Yates, D.C.; Mitcheson, P.D. Modelling and analysis of Class EF and Class E/F inverters with series-tuned resonant networks. *IEEE Trans. Power Electron.* **2016**, *31*, 3415–3430. [CrossRef]
13. Mugisho, M.S.; Thian, M.; Grebennikov, A. Analysis and Design of a High-Efficiency Class-EM Power Amplifier. In Proceedings of the 2019 IEEE Radio and Wireless Symposium (RWS), Orlando, FL, USA, 20–23 January 2019; pp. 1–4.

14. Beltran, R.A.; Raab, F.H. Simplified analysis and design of outphasing transmitters using class-E power amplifiers. In Proceedings of the 2015 IEEE Topical Conference on Power Amplifiers for Wireless and Radio Applications (PAWR), San Diego, CA, USA, 25–28 January 2015; pp. 1–3.
15. Chadha, A.; Ayachit, A.; Saini, D.K.; Kazimierczuk, M.K. Steady-state analysis of PWM tapped-inductor buck DC-DC converter in CCM. In Proceedings of the 2018 IEEE Texas Power and Energy Conference (TPEC), College Station, TX, USA, 8–9 February 2018; pp. 1–6.
16. Suetsugu, T.; Kazimierczuk, M. Comparison of class-e amplifier with nonlinear and linear shunt capacitance. *IEEE Trans. Circuits Syst. I Fundam. Theory Appl.* **2003**, *50*, 1089–1097. [CrossRef]
17. Zhang, Z.; Lin, J.; Zhou, Y.; Ren, X. Analysis and Decoupling Design of a 30 MHz Resonant SEPIC Converter. *IEEE Trans. Power Electron.* **2016**, *31*, 4536–4548. [CrossRef]
18. Guan, J.; Negra, R. Steady-state analysis and fast optimisation of Class-E power amplifiers with lossy switch for RF choke and finite DC-feed inductance. In Proceedings of the 2013 IEEE 56th International Midwest Symposium on Circuits and Systems (MWSCAS), Columbus, OH, USA, 4–7 August 2013; pp. 380–383.
19. Reynaert, P.; Mertens, K.; Steyaert, M. A state-space behavioral model for CMOS class E power amplifiers. *IEEE Trans. Comput. Aided Des. Integr. Circuits Syst.* **2003**, *22*, 132–138. [CrossRef]
20. Liang, J.; Liao, W. Steady-State Simulation and Optimization of Class-E Power Amplifiers With Extended Impedance Method. *IEEE Trans. Circuits Syst. I Regul. Pap.* **2011**, *58*, 1433–1445. [CrossRef]
21. Zhou, J.; Morris, K.A.; Watkins, G.T.; Yamaguchi, K. Improved Reactance-Compensation Technique for the Design of Wideband Suboptimum Class-E Power Amplifiers. *IEEE Trans. Microw. Theory Tech.* **2015**, *63*, 2793–2801. [CrossRef]
22. Khansalee, E.; Nuanyai, K.; Zhao, Y. Design and implementation of class E power amplifier with parallel circuit for wireless power transfer systems. In Proceedings of the 2017 International Electrical Engineering Congress (iEECON), Pattaya, Thailand, 8–10 March 2017; pp. 1–4.

© 2019 by the authors. Licensee MDPI, Basel, Switzerland. This article is an open access article distributed under the terms and conditions of the Creative Commons Attribution (CC BY) license (http://creativecommons.org/licenses/by/4.0/).

Article

Optimal Design and Comparison of High-Frequency Resonant and Non-Resonant Rotary Transformers [†]

Koen Bastiaens *, Dave C. J. Krop, Sultan Jumayev and Elena A. Lomonova

Department of Electrical Engineering, Electromechanics and Power Electronics, Eindhoven University of Technology, 5600 MB Eindhoven, The Netherlands; d.c.j.krop@tue.nl (D.C.J.K.); s.jumayev@tue.nl (S.J.); e.lomonova@tue.nl (E.A.L.)
* Correspondence: k.bastiaens@tue.nl
† This paper is an extended version of our paper published in Bastiaens, K.; Krop, D.C.J.; Jumayev, S.; Lomonova, E.A. Design and Comparison of High-Frequency Resonant and Non-Resonant Rotating Transformers. In Proceedings of the 21st International Conference on Electrical Machines and Systems (ICEMS), Jeju, South-Korea, October 2018; pp. 1703–1708.

Received: 20 January 2020; Accepted: 17 February 2020; Published: 19 February 2020

Abstract: This paper concerns the optimal design and comparative analysis of resonant and non-resonant high-frequency GaN-based rotating transformers. A multi-physical design approach is employed, in which magnetic, electrical, and thermal models are coupled. The results are verified by experiments. Two different optimization objectives are considered; firstly, the efficiency of two standard core geometries is maximized for a required output power level. Secondly, a geometrical optimization is performed, such that the core inertia is minimized for the desired output power level. The results of both design optimizations have shown large improvements in terms of output power and core inertia as a result of applying series–series resonant compensation.

Keywords: design optimization; finite element analysis; gallium nitride; gradient methods; inductive power transmission; power measurement; transformer cores

1. Introduction

Wireless power transfer (WPT) is widely employed in applications that require reliable power transfer to rotating parts, e.g., in battery charging and robotic applications [1,2], as well as an alternative to slip rings or brushes in electric machines [3,4]. Generally, a cylindrical transformer is used, which has a rotary and stationary side separated by a small air gap in either the radial or axial direction. Figure 1 shows an example of such a cylindrical transformer geometry, specifically a pot core transformer. A high magnetic coupling is achieved by the application of core material that has a high permeability [5]. Furthermore, a high-frequency power supply is typically applied, such that the transfer of power and efficiency are enhanced, whereas the volume of the transformer is reduced [6]. Gallium-Nitride (GaN) transistors have gained popularity in WPT applications, since switching frequencies in the range of several megahertz (MHz) are realized. Moreover, in comparison to conventional Silicon devices, the switching losses are lower, whereas the power density is higher [7]. Additionally, the leakage inductance is often compensated by the application of resonant techniques. Therefore, capacitors are placed parallel to, or in series with, the transformer winding on either or both the primary and secondary side, in doing so enhancing the transfer of power [8]. In low-voltage systems resonant techniques might be undesirable, since high voltages across the capacitors can occur [9].

The optimal design of a high-frequency rotary WPT system can be challenging, since different physical domains (i.e., magnetic, electrical, and thermal) and various design parameters have to be considered. Several design studies and methodologies are available in literature [10–16]. The design and analysis are often performed for a fixed electrical frequency and core geometry [10,11]. Furthermore, typically either series–series resonant compensation [10,12,13] or non-resonant systems are considered [11,14]. In [10], a WPT system based on a pot core transformer was designed and a prototype was realized, which at an electrical frequency of 50 kHz achieved an output power of 1.3 kW. The multi-physical design approach that was applied in this work, consists of equivalent circuit analysis for the electrical, magnetic, and thermal domains. Series–series resonance was applied in the design approach and prototype. In [11], a comparable WPT system was designed for the same electrical frequency. The design approach consists of an electrical circuit model simulated in commercial software (i.e., MATLAB Simulink), a magnetic equivalent circuit model, and the thermal model applies a Finite Element Method (FEM) model. Resonant techniques were not considered in this work. Alternatively, design studies often consider a limited or low frequency range [12–14]. In [12], two different winding topologies for a rotating pot core transformer were compared in terms of core volume and power losses. A design optimization was performed for frequencies in the range of 1 kHz–200 kHz. The multi-physical design approach proposed in this paper is based on equivalent circuit models for the electrical, magnetic, and thermal domain. Experimental results were obtained for both winding configurations. Series–series resonance was applied in the design approach and prototype. In [13], three different winding topologies for a rotating WPT system were proposed and the performance was measured for varying frequencies in the range of 440 kHz–612 kHz. A GaN half-bridge inverter and series–series resonance were applied in the system. A maximum output power of 20 W and an efficiency of 89.7% were realized. In [14], two different non-resonant cylindrical transformer topologies were compared by means of a design optimization. The multi-physical design model is based on equivalent circuit models for both the magnetic and thermal domains, whereas the electrical circuit model is simulated using commercial software (i.e., MATLAB SimPowerSystem Toolbox). Furthermore, design studies often only investigate the behavior of the system in the magnetic and electrical domains [13,15,16]. In [15], a three-phase rotary WPT system was designed by means of a magnetic equivalent circuit model. A similar system was designed in [16], in this work the design was carried out using a FEM magnetic model. However, the optimal design is also influenced by the temperature distribution and the corresponding constraints. Multi-physical design approaches are discussed in literature [10–12]. These design methodologies typically apply magnetic equivalent circuit models [10–12] and thermal equivalent circuit models [10,12], which are capable of realizing computationally efficient design routines. Equivalent circuit models provide adequate accuracy in most cases, however compared to the most commonly applied numerical method, i.e., the FEM, the accuracy is generally lower [17]. Therefore, research on a full system approach, in which all physical domains are covered (i.e., magnetic, electrical, and thermal) and a wide design space is investigated (i.e., geometrical, frequency, and both series–series and non-resonant circuits) is lacking.

In this paper, a multi-physical design approach for the optimal design of both resonant and non-resonant high-frequency rotary transformers is presented. The design approach couples both a magnetic and thermal FEM model as well as a Spice electrical circuit model. The design approach can be applied to any arbitrary objective function and rotary transformer topology, in this paper the design approach is applied to two different optimization problems. First, the efficiency of two fixed pot core geometries is maximized for a desired output power level of at least 100 W. A frequency range up to and including 1 MHz is considered, by the application of a GaN half-bridge inverter. Second, the core inertia is minimized for the desired output power level. In both cases a comparative analysis of the resonant and non-resonant designs is performed. The optimal design resulting from the first optimization problem is prototyped for the purpose of experimental verification.

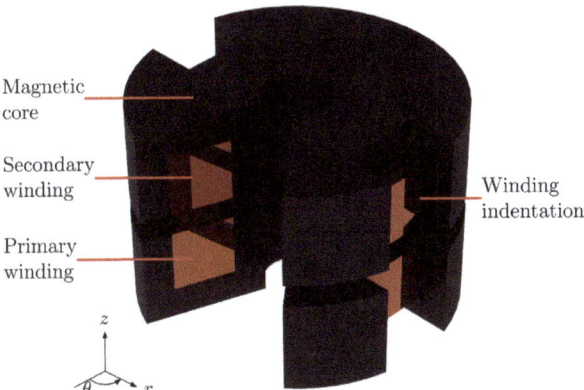

Figure 1. Wedge-shaped cross-sectional view of an axially gapped cylindrical (pot core) transformer, including indication of the various components.

2. Materials and Methods

2.1. Wireless Power Transfer System

The high-frequency WPT system under consideration in this paper consists of a 48 V_{dc} supply, a dc/ac half-bridge converter, the WPT coils (each of which is placed inside a magnetic pot core separated by an air gap), and a single-phase rectifier including load. The pot core geometry, as shown in Figure 1, is selected to be the rotating transformer topology under investigation. In the pot core transformer, the power is transferred in the axial direction, thus is referred to as the axially gapped topology. Alternatively, the power can also be transferred in the radial direction, by changing the air gap configuration, i.e., radially gapped topology. However, with respect to the radially gapped topology, the axially gapped topology is favorable in terms of magnetic coupling and losses [14]. Furthermore, each coil is placed in a separate core half, as shown in Figure 1, which is typically referred to as the adjacent coil configuration. Alternatively, the coils can be arranged in the coaxial configuration, in which one coil rotates inside the other. However, the adjacent configuration is favorable in terms of losses [12]. The high-frequency WPT system is designed by the application of coupled magnetic, electrical, and thermal models.

2.2. Magnetic Model

The magnetic model employs a two-dimensional FEM steady state ac model in the axisymmetrical plane, which is solved by commercial software, i.e., Altair Flux [18]. The modeled geometry is shown in Figure 2. The magnetic model is used to calculate the magnetizing and leakage inductances, which serve as inputs to the electrical model. The magnetizing and leakage inductances are calculated as:

$$L_{lkp} = L_p - \left(\frac{N_p}{N_s}\right) M, \tag{1}$$

$$L_{lks} = L_s - \left(\frac{N_s}{N_p}\right) M, \tag{2}$$

$$L_m = \left(\frac{N_p}{N_s}\right) M, \tag{3}$$

where L_{lk} is the leakage inductance, L is the self inductance, N is the number of turns, M is the mutual inductance, L_m is the magnetizing inductance, and the subscripts p and s are used to indicate the primary and secondary side, respectively. Under the assumption that the transformer is magnetically linear, the self- and mutual inductances are calculated from the apparent inductance according to:

$$L_p = \frac{N_p \phi_p}{I_p}, \tag{4}$$

$$L_s = \frac{N_s \phi_s}{I_s}, \tag{5}$$

$$M_{ps} = \frac{N_s \phi_s}{I_p} = M_{sp} = \frac{N_p \phi_p}{I_s} = M, \tag{6}$$

where I_p and I_s are the root-mean-square (rms) values of the primary and secondary current, respectively, ϕ_p and ϕ_s are the flux in the primary and secondary coil, respectively (obtained from the FEM model), M_{ps} is the mutual inductance between the primary and secondary coil, and M_{sp} is the mutual inductance between the secondary and primary coil. The magnetic coupling coefficient (k) is obtained from the mutual and self inductances according to:

$$k = \frac{M_{ps}}{\sqrt{L_p L_s}}. \tag{7}$$

The magnetic coupling coefficient represents the degree of magnetic coupling, thus a coefficient equal to one represents perfect coupling (i.e., zero leakage inductance) [19].

Additionally, the magnetic model is employed for the calculation of the iron losses in the transformer core. Therefore, both sides of the core geometry are divided into five regions, as shown in Figure 2. In every region the iron losses (P_{Fe}) are calculated, which serve as inputs to the thermal model and efficiency calculation. The iron losses are calculated according to Steinmetz's equation, given by:

$$P_{Fe} = \int_{V_i} C_m f^\alpha B_i^\beta dV_i, \tag{8}$$

where C_m [W·s$^\alpha$/T$^\beta$/m^3], α [-], and β [-] are empirical parameters, which in this case are set to 10.6, 1.3, and 2.7, respectively [20], f is the electrical frequency, B_i and V_i are the magnetic flux density and volume, respectively of the corresponding region i [21]. Steinmetz's equation is valid for sinusoidal excitation, which has been assumed in the magnetic model. However, for the non-resonant transformer, the half-bridge converter induces non-sinusoidal currents. Consequently, discrepancies in the iron losses and peak magnetic flux density are introduced. However, as a result of the Joule losses being dominant with respect to the iron losses, the effect of this assumption on the temperature rise and efficiency is negligible.

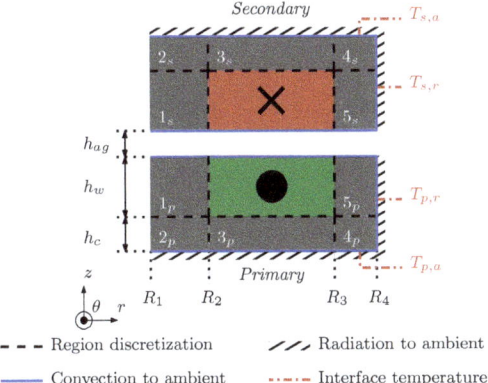

Figure 2. Two-dimensional representation of the pot core transformer geometry in the axisymmetrical plane, including indication of geometrical parameters, region discretization, thermal boundaries, and interface temperatures.

2.3. Electrical Model

The electrical equivalent circuit model of the WPT system is shown in Figure 3. The circuit model is based on the transformer T-model, which consists of the magnetizing inductance, leakage inductances, and the coil resistances [19]. The contribution of the core losses is included in the efficiency calculation, however the effect of the iron losses is assumed to be negligible in the electrical equivalent circuit model. The components on the secondary side of the circuit, are reflected to the primary side through the winding ratio according to:

$$Z' = Z \left(\frac{N_p}{N_s}\right)^2, \tag{9}$$

where Z is the impedance on the secondary side, and Z' is the reflected impedance [19]. Both non-resonant and series–series resonant electrical circuits, which is the most commonly used technique, are considered in the model. The primary and secondary capacitance, which are required to compensate the leakage inductances, are calculated according to:

$$C_p = \frac{1}{(2\pi f)^2 L_{lkp}}, \tag{10}$$

$$C'_s = \frac{1}{(2\pi f)^2 L'_{lks}}, \tag{11}$$

where C_p and C'_s are the primary and secondary capacitances, respectively.

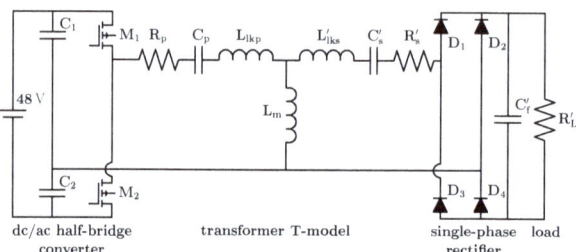

Figure 3. Electrical equivalent circuit model including series–series resonant capacitors and indication of the various components.

The dc/ac half-bridge converter selected in the circuit model is the EPC9035 development board from Efficient Power Conversion Corporation [22], which employs GaN transistors, such that frequencies up to and including 1 MHz can be investigated. In order to account for the high electrical frequencies, PMEG6030ETP Schottky diodes from Nexperia [23], which have a short reverse recovery time of 12 ns, are used in the single-phase rectifier. Additionally, in order to improve the power dissipation in the single-phase rectifier and reduce the thermal load per diode, two diodes are placed in parallel.

Litz wire is employed in the transformer windings, such that the losses caused by the skin- and proximity-effect are minimized. The additional losses caused by these effects are approximated in the design model by the ac resistance (R_{ac}), which is obtained by scaling the dc resistance according to:

$$R_{ac} = R_{dc} \left(\frac{1}{k_s}\right) \left(1 + \frac{\beta_R^2}{192 + \beta_R^2}\right), \tag{12}$$

$$k_s = \begin{cases} 1, & \text{if } \delta \geq r_s, \\ \frac{r_s^2 - (r_s^2 - \delta)^2}{r_s^2}, & \text{otherwise,} \end{cases} \tag{13}$$

$$\beta_R = \frac{2\mu_0 f}{R_{dc}}, \tag{14}$$

$$R_{dc} = \rho \frac{l_w}{A_w}, \tag{15}$$

where R_{dc} is the dc resistance of the wire, δ is the skin-depth, r_s is the strand radius [24–26], ρ is the resistivity of copper, l_w is the length of the wire, and A_w is the effective copper area of the wire. The parameters of the litz wire, from Pack Litz Wire, considered in the design analysis are shown in Table 1 [27]. In order to minimize the skin-effect losses, the strand diameter of the litz wire is chosen such that it is always smaller than the skin depth. As a result of the litz wire having a relatively small outer diameter, the fill-factor (the ratio between the effective copper area and total available winding area) is maximized by placing wires in parallel. The maximum fill-factor is assumed to be equal to 0.16 and in case wires are placed in parallel, the maximum fill-factor is reduced to 0.13, since the wires are braided in order to minimize the proximity effect losses.

Table 1. Litz wire parameters.

Parameter	Symbol	Value	Unit
Strand radius	r_s	0.016	mm
Number of strands	n_s	250	-
Nominal dc resistance (20 °C)	$R_{dc,0}$	0.0987	Ω/m
Total wire diameter including wrapping	d_w	0.52	mm

The electrical circuit model is solved using the LT-Spice circuit simulator [28], in which the non-ideal device models of the Schottky diodes and GaN transistors are included [23,29]. Finally, once the model has reached a steady-state, the input power (P_{in}), output power (P_o), efficiency (η), and power losses in the coils (P_c), switches (P_{sw}) and diodes (P_d) are extracted from the model according to:

$$P_{in} = V_{dc} I_{dc}, \tag{16}$$

$$P_o = \frac{1}{T} \int_0^T v_o(t) i_o(t) \, dt, \tag{17}$$

$$\eta = \frac{P_o}{P_{in} + P_{Fe}} \times 100\%, \tag{18}$$

$$P_c = \frac{1}{T} \int_0^T \left(i_p(t)^2 R_{p,ac} + i_s(t)^2 R_{s,ac} \right) dt, \tag{19}$$

$$P_{sw} = \sum_{i=1}^{N_{sw}} \frac{1}{T} \int_0^T \left(v_{DS,i}(t) i_{D,i}(t) + v_{GS,i}(t) i_{G,i}(t) \right) dt, \tag{20}$$

$$P_d = \sum_{i=1}^{N_d} \frac{1}{T} \int_0^T v_{d,i}(t) i_{d,i}(t) \, dt, \tag{21}$$

where V_{dc} and I_{dc} are the dc bus voltage and current, respectively, T is the time period, $v_o(t)$ and $i_o(t)$ are the output current and voltage, respectively, $i_p(t)$ and $i_s(t)$ are the primary and secondary current, respectively, $R_{p,ac}$ and $R_{s,ac}$ are the ac resistances of the primary and secondary coil, respectively, N_{sw} is the number of switches, $v_{Ds,i}(t)$ and $i_{D,i}(t)$ are the drain-to-source voltage and drain current in switch i, respectively, $v_{GS,i}(t)$ and $i_{G,i}(t)$ are the gate-to-source voltage and gate current, respectively, N_d is the number of diodes, $v_{d,i}(t)$ and $i_{d,i}(t)$ are the voltage and current across diode i. The Joule losses in the coils serve as an input to the thermal model.

2.4. Thermal Model

As a result of the Joule losses in the windings and the iron losses in the core, a temperature rise is generated in the magnetic pot core. In order to approximate this temperature rise, a two-dimensional FEM thermal model is employed. The thermal model includes the heat transfer by means of conduction between the various regions. The heat transfer through convection and radiation to the ambient environment is incorporated on the boundaries of the geometry, where the former is also included in the air gap region. The shaft or mounting point, which is typically present at the inner radial boundary, is assumed to have the same thermal properties as air. Furthermore, in order to evaluate a worst-case scenario, the effect of rotation is neglected. The thermal boundary conditions are included in Figure 2. The emissivity coefficient is assumed to be equal to 0.31, which is a typical value for dark-gray iron surfaces [30]. The convection coefficient (h) is given by:

$$h = \frac{\bar{N}_u k_a}{X}, \tag{22}$$

where k_a is the thermal conductivity of the ambient air, X is the characteristic length (given by $1.8R_o$ and $2R_o$ at the axial and radial boundaries, respectively), and $\tilde{N}_u$ is the overall Nusselt number, which is given by:

$$\tilde{N}_u = \begin{cases} C_1(G_rP_r)^{\frac{1}{4}}, & \text{if } G_r \leq 10^5 \text{ (laminar flow)}, \\ C_2(G_rP_r)^{\frac{1}{3}}, & \text{otherwise (turbulent flow)}, \end{cases} \quad (23)$$

where G_r and P_r are the Grashof and Prandtl number, respectively, C_1 and C_2 are empirical coefficients, which at the radial boundaries are set to 0.47 and 0.10, respectively, and at the axial boundaries the coefficients are set to 0.54 and 0.14, respectively. The Grashof number is dependent on the temperature at the interface (i.e., the axial and radial interfaces on both the primary and secondary core; $T_{p,a}$, $T_{p,r}$, $T_{s,a}$, $T_{s,r}$) and is given by:

$$G_r = \frac{g\beta \left(T_i - T_\infty\right) X^3}{v_a^2}, \quad (24)$$

where g is the gravitational acceleration, β is the coefficient of thermal expansion (assuming an ideal gas; $\beta = T_\infty^{-1}$), T_i is the interface temperature, T_∞ is the temperature of the ambient air, and v_a is the kinematic viscosity of the ambient air [31]. Furthermore, the resistivity of the copper is a function of the coil temperature, given by:

$$\rho(T_c) = \rho_0 \left(1 + \alpha \left(T_c - T_0\right)\right), \quad (25)$$

where ρ_0 is the resistivity at temperature T_0, T_c is the coil temperature, and α is the temperature coefficient [32]. As a result of the temperature dependence shown in (24) and (25), the thermal model is solved in an iterative manner, recalculating the coil resistances and heat transfer coefficients at every iteration until a steady-state is reached.

2.5. Pot Core Design Optimization

A design optimization is performed, in which the standard P14/8 [20] and P18/11 [20] pot core geometries, from Fair-Rite Products, for both non-resonant and resonant compensation are considered. The volume of a core half and inertia of the P14/8 pot core are equal to 0.365 cm^3 and 49.3 g·mm^2, respectively, whereas for the P18/11 pot core the quantities are equal to 0.821 cm^3 and 181.6 g·mm^2, respectively. The resulting optimal design is constructed, and measurements are performed in order to verify the design models and results. The objective of the design optimization is to maximize the efficiency for an output power level of at least 100 W, whilst satisfying the constraints. The optimization problem is given by:

$$\begin{aligned} \underset{\vec{x}}{\text{maximize:}} \quad & \eta(\vec{x}), \\ \text{where:} \quad & \vec{x} = \{f, N_p, N_s\}, \\ \text{subject to:} \quad & P_o(\vec{x}) \geq 100 \text{ W}, \\ & \hat{B}_i(\vec{x}) \leq 350 \text{ mT}, \\ & T_i(\vec{x}) \leq 100\,^\circ\text{C}, \\ & T_w(\vec{x}) \leq 150\,^\circ\text{C}, \\ & I_p(\vec{x}) \leq 25.0 \text{ A}, \\ & I_s(\vec{x}) \leq 3.00 \text{ A}, \\ & V_{c,i}(\vec{x}) \leq 48.0 \text{ V}, \end{aligned} \quad (26)$$

where η is the efficiency, $\vec{x}$ is the set of design variables, which consists of the electrical frequency (f), the number of primary (N_p) and secondary turns (N_s), respectively, P_o is the output power, $\hat{B}_i$ and T_i are the maximum value of the magnetic flux density and average value of the temperature in the various core regions, T_w is the average winding temperature, I_p and I_s are the rms-values of the primary and secondary current, respectively, and $V_{c,i}$ is the rms-value of the voltage across the resonant capacitors. The ambient temperature is assumed to be equal to 20 °C. Furthermore, the axial height of the air gap that separates the primary and secondary core (h_{ag}) and load resistance are fixed to 0.5 mm and 23.5 Ω, respectively.

The optimization problem, shown in (26), is solved by applying the parametric search method. The frequency is varied from 50 kHz up to and including 1 MHz in incremental steps of 25 kHz. For the number of turns, incremental steps of one turn are taken in the range starting at one and stopping at the point where the maximum allowable fill factor is exceeded. At every iteration, the design closest to the output power constraint and satisfying all other constraints is stored.

2.6. Experimental Verification

For the purpose of experimentally verifying the design approach, a stationary prototype of the optimal design resulting from (26) is realized. The prototype was used for the measurement of the input and output quantities. Furthermore, the core temperatures were measured by thermocouples fixed to the axial and radial interfaces with the ambient air. The thermocouples were mounted on both the primary and secondary side of the pot core. The various interface temperatures (T) are indicated in red in Figure 2, where the subscripts p and s denote the primary and secondary side, respectively, while the subscripts a and r are used to mark the axial and radial interface, respectively. The measured quantities are compared to the simulation results.

2.7. Geometrical Optimization

Alternatively to optimizing a fixed core geometry, as shown in (26), the required core inertia for realizing the desired output power level can also be minimized within the investigated frequency range. In this situation, the optimization problem is given by:

$$\begin{aligned}
&\underset{\vec{x}}{\text{minimize:}} && J_c(\vec{x}), \\
&\text{where:} && \vec{x} = \{f, N_p, N_s, R_2, R_3, h_w\}, \\
&\text{subject to:} && P_o(\vec{x}) \geq 100 \text{ W}, \\
& && h_c = \frac{R_2^2 - R_1^2}{R_3 + R_2}, \\
& && R_4 = \sqrt{R_2^2 - R_1^2 + R_3^2}, \\
& && \{h_c, R_2 - R_1, R_4 - R_3\} \geq 1.00 \text{ mm}, \\
& && \{h_w, R_3 - R_2\} \geq 2.00 \text{ mm},
\end{aligned} \quad (27)$$

where J_c is the inertia of the core, excluding the winding, R_2 and R_3 are the inner and outer radius of the winding area, respectively, and h_w is the axial height of the winding area. The inertia of the core is calculated as:

$$\begin{aligned}
J_c =& \frac{1}{2}\pi (h_c + h_w) \rho_m \left(R_2^4 - R_1^4\right) + \frac{1}{2}\pi h_c \rho_m \left(R_3^4 - R_2^4\right) \\
&+ \frac{1}{2}\pi (h_c + h_w) \rho_m \left(R_4^4 - R_3^4\right),
\end{aligned} \quad (28)$$

where ρ_m is the mass density of the ferrite, which in this case is equal to 4800 kg/m^3 [20]. For the sake of reducing the number of design variables, the axial height of the bottom core part (h_c) is calculated in such a way that, the cross-sectional area of region 3 at the average radius is equal to the cross-sectional area of region 1. The minimum axial height of the bottom core part (h_c) is equal to 1.0 mm. The outer radius of the core (R_4) is calculated such that the cross-sectional areas of regions 1 and 5 are equal to each other, and the radial depth of region 5 is at least equal to 1.0 mm, which is the minimum radial depth for all regions. Consequently, an approximately equal magnetic flux density is obtained in regions 1, 3, and 5. The inner radius of the core (R_1) is fixed to 1.60 mm, which is the same value as for the P18/11 pot core. The minimum axial height and radial thickness of the winding area are equal to 2.0 mm, such that sufficient space for fitting and gluing the winding is realized. The core regions and geometrical variables are indicated in Figure 2. The constraints on the maximum value of the magnetic flux density, average value of the core and winding temperature, rms-values of the primary and secondary current, and rms-values of the voltage across the resonant capacitors remain equal to the constraints shown in (26) and are therefore not repeated in (27). Additionally, the ambient temperature, air gap height and load resistance are unchanged.

The optimization problem is solved by a gradient-based algorithm (i.e., interior-point from MATLAB [33]) for five different initial points, where the first initial point is the P18/11 pot core geometry and the other four are generated at random (by means of the Multistart algorithm from MATLAB [33]). Within the optimization problem, again both non-resonant and resonant compensation are considered. The winding ratio is determined in an internal parametric search loop, as these variables are discrete. In this internal loop, the primary and secondary number of turns are incrementally changed in order find the combination closest to the output power constraint, while respecting the maximum copper fill factor.

3. Results

3.1. Pot Core Design Optimization Results

The resulting output power as a function of the electrical frequency for both core geometries, non-resonant, and resonant compensation are shown in Figure 4. From the results, the average increase in output power by applying series–series resonance is calculated according to:

$$\Delta P = \overline{\left(\frac{P_r(f) - P_{nr}(f)}{P_{nr}(f)}\right)} \times 100\%, \tag{29}$$

where P_r and P_{nr} are the output power for the resonant and non-resonant designs as a function of the frequency (f), respectively. Equation (29) is only evaluated for the frequencies at which both a resonant and non-resonant transformer design is obtained, i.e., the frequency ranges 275 kHz–1 MHz and 125 kHz–1 MHz for the P14/8 and P18/11 cores, respectively. Consequently, as a result of applying series–series resonance within the optimization problem, an average increase in output power of 39.7% and 45.5% is observed for the P14/8 and P18/11 pot cores, respectively. For electrical frequencies below the evaluated ranges, the constraint on the peak magnetic flux density is not satisfied.

The constraint on the output power (at least 100 W of output power) is only realized by the P18/11 pot core in combination with resonant compensation within the frequency range of 325 kHz–1 MHz. The efficiency corresponding to the feasible frequency range is shown in Figure 5. The optimum is located at an electrical frequency of 500 kHz, at which the overall system efficiency is equal to 92.8%. As a result of the discrete step in the winding ratio, and the rounding of the resonant capacitances, the output power and efficiency as function of frequency characteristics have non-smooth behavior.

The separation of the power losses into the various components (i.e., Joule, iron, switch, and diode losses, respectively) for the optimal transformer design is shown in Figure 6. The largest portion of the losses occurs in the diodes, whereas the iron losses give a negligible contribution (approximately equal to 0.1 W). Additionally, the figure demonstrates the high efficient operation of the GaN half-bridge inverter. The ratio between the Joule and the iron losses for the optimal transformer design is approximately thirteen, therefore the assumption that has been made in the magnetic model on the dominance of the Joule losses is validated.

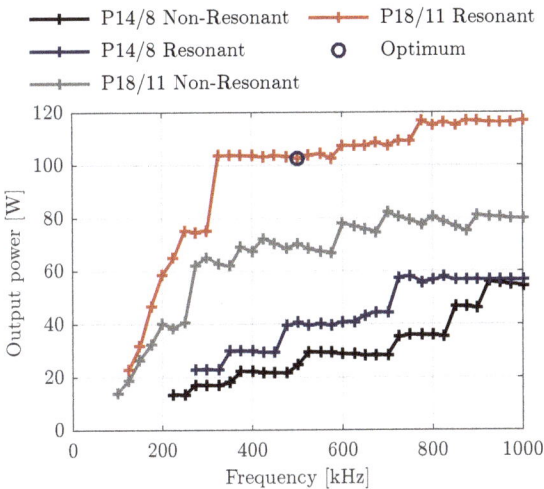

Figure 4. Optimization results: Output power as a function of frequency for all designs.

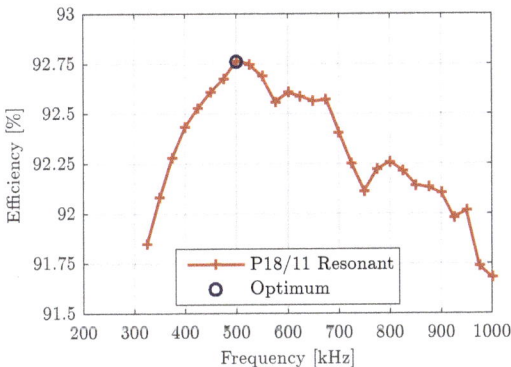

Figure 5. Optimization results: Efficiency as a function of frequency for the P18/11 resonant transformer in the feasible output power region.

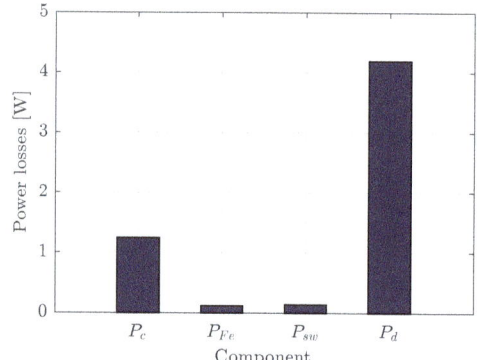

Figure 6. Optimization results: Separation of power losses for the optimal P18/11 resonant pot core design; Joule (P_c), iron (P_{Fe}), switch (P_{sw}), and diode losses (P_d), respectively.

3.2. Experimental Results

For the purpose of experimental verification of the design approach, a stationary prototype of the optimal P18/11 pot core resonant transformer design is realized. The corresponding transformer parameters resulting from the simulation are shown in Table 2. The measurements were performed at an ambient temperature of 22 °C and the dc bus voltage was set to 48 V.

The output current and voltage as a function of time resulting from both the measurements and the simulation are shown in Figure 7. The corresponding average values (I_o and V_o, respectively) are shown in Table 3. Additionally, the discrepancy (ϵ) between the simulation and measurement results is included, which is calculated as:

$$\epsilon = \frac{f_s - f_m}{f_m} \times 100\%, \tag{30}$$

where f_s and f_m represent the values obtained by the simulation and measurement, respectively. A good agreement between the simulation and measurement results is achieved, small discrepancies of -0.95% and $+1.9\%$ are observed in the average values of the output current and voltage, respectively. Consequently, the transferred power resulting from the simulation closely matches the measurements, i.e., a small discrepancy of $+0.49\%$ is observed. Furthermore, low discrepancies of -3.1% and $+3.1\%$ are observed in the input power and efficiency calculations, respectively.

Additionally, the measured and estimated temperatures are shown in Table 3. On both sides a good agreement between the simulation results and the measurements is observed, small discrepancies of $+2.1\%$ and $+5.4\%$ are observed in the axial ($T_{s,a}$) and radial ($T_{s,r}$) interface temperatures on the secondary side, respectively. On the primary side, higher discrepancies are observed; $+8.2\%$ and $+10.1\%$ for the axial ($T_{p,a}$) and radial ($T_{p,r}$) interfaces, respectively. The higher discrepancy on the primary side might be caused by unaccounted heat transfer from the core to the printed circuit board (PCB). However, the model realizes sufficient accuracy for the design of the WPT system, therefore providing validation of the thermal model.

Table 2. Optimal P18/11 pot core resonant transformer design.

Parameter	Symbol	Value	Unit
Frequency	f	500	kHz
Number of primary turns	N_p	5	-
Number of secondary turns	N_s	11	-
Number of parallel paths primary side	a_p	2	-
Number of parallel paths secondary side	a_s	1	-
Magnetizing inductance	L_m	2.125	µH
Primary leakage inductance	L_{lkp}	430.4	nH
Secondary leakage inductance	L'_{lks}	430.4	nH
Magnetic coupling coefficient	k	0.893	-
Primary resonant capacitance	C_p	240.0	nF
Secondary resonant capacitance	C'_s	237.2	nF

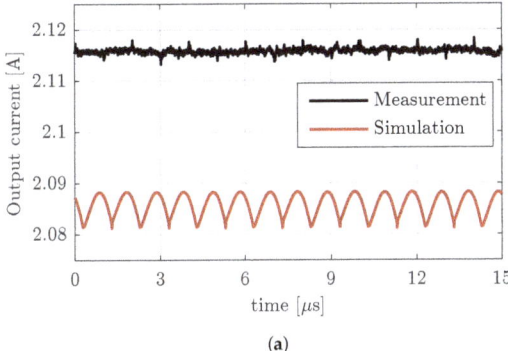

(a)

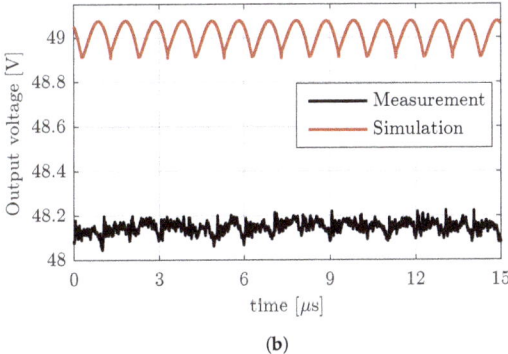

(b)

Figure 7. Measurement and simulation results: (**a**) Output current as a function of time, and (**b**) output voltage as a function of time.

Table 3. Comparison of simulation and measurement results.

Quantity	Symbol	Measurement	Simulation	Unit	Discrepancy [%]
Output current	I_o	2.11	2.09	A	−0.95
Output voltage	V_o	48.1	49.0	V	+1.9
Output power	P_o	101.7	102.2	W	+0.49
Input power	P_{in}	113.6	110.1	W	−3.1
Efficiency	η	89.6	92.8	%	+3.1
Secondary core axial interface temperature	$T_{s,a}$	72.5	74.0	°C	+2.1
Secondary core radial interface temperature	$T_{s,r}$	69.9	73.7	°C	+5.4
Primary core axial interface temperature	$T_{p,a}$	76.5	82.8	°C	+8.2
Primary core radial interface temperature	$T_{p,r}$	73.9	81.4	°C	+10.1

3.3. Geometrical Optimization Results

The two resulting optimal transformer geometries are shown in Figure 8. The corresponding geometrical parameters, transformer designs, and physical quantities are shown in Table 4. Compared to the previously determined P18/11 pot core design, a reduction of the core inertia by 38.2% and 66.4% are realized by the optimal non-resonant and resonant design, respectively. In order to obtain the reduction in core inertia, a higher electrical frequency is utilized (i.e., 850 and 950 kHz for the non-resonant and resonant design, respectively compared to 500 kHz for the optimal P18/11 design), which results in increased losses in the power electronics. Consequently, with respect to the P18/11 design, the non-resonant and resonant design decrease the efficiency by 1.5% and 1.1%, respectively.

Compared to the non-resonant optimal design, the resonant optimal design reduces the core inertia by 45.5%, while the output power and efficiency are increased by 9.3% and 0.4%, respectively. The increase in output power above 100 W by the resonant design is caused by the discrete step in the winding design.

In case of the non-resonant design, the core inertia is minimized through the minimization of the leakage inductances. Consequently, the winding area has a small axial height and a large radial width, as shown in Figure 8a. As a result, a high magnetic coupling factor of 0.95 is achieved for the non-resonant design. For the resonant design, the leakage inductances are compensated. Consequently, the opposite is true; the minimization of the core inertia is realized by minimizing the radial width, while utilizing a larger axial height, as shown in Figure 8b. Consequently, with respect to the non-resonant design, a lower magnetic coupling factor is obtained, since the leakage inductances are higher. Furthermore, a higher number of turns is fitted, such that a high magnetizing inductance is created.

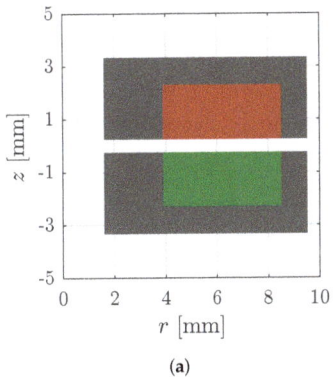
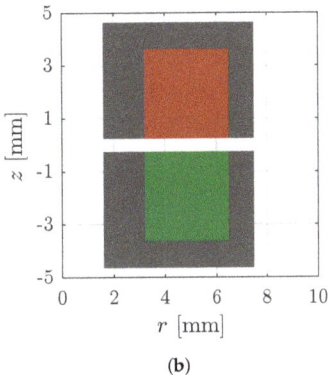

(a)   (b)

Figure 8. Optimized transformer geometries: (**a**) Non-resonant and (**b**) resonant transformer design.

Table 4. Optimal non-resonant and resonant transformer designs resulting from the geometrical optimization.

Parameter	Symbol	Non-Resonant	Resonant	Unit
Geometrical Design				
Inner radius core	R_1	1.60	1.60	mm
Inner radius winding area	R_2	3.91	3.23	mm
Outer radius winding area	R_3	8.52	6.48	mm
Outer radius core	R_4	9.52	7.48	mm
Winding area height	h_w	2.05	3.38	mm
Height bottom core part	h_c	1.02	1.00	mm
Core inertia	J_c	112.2	61.1	g·mm^2
Core volume	V_c	0.481	0.400	cm^3
Transformer Design				
Frequency	f	850	950	kHz
Number of primary turns	N_p	3	4	-
Number of secondary turns	N_s	8	9	-
Number of parallel paths primary side	a_p	2	2	-
Number of parallel paths secondary side	a_s	1	1	-
Magnetizing inductance	L_m	0.801	0.974	µH
Primary leakage inductance	L_{lkp}	86.3	248.0	nH
Secondary leakage inductance	L'_{lks}	86.3	248.0	nH
Magnetic coupling coefficient	k	0.949	0.888	-
Primary resonant capacitance	C_p	-	110.0	nF
Secondary resonant capacitance	C'_s	-	111.4	nF
Physical Quantities				
Quantity	Symbol	Non-Resonant	Resonant	Unit
Output power	P_o	100.6	110.0	W
Efficiency	η	91.3	91.7	%
Max. temperature primary core	$\hat{T}_{p,i}$	94.6	95.1	°C
Max. temperature secondary core	$\hat{T}_{s,i}$	78.7	84.4	°C

4. Discussion

4.1. Three-Dimensional Effects

In each side of the pot core geometry, a small indentation is present, as shown in Figure 1, such that the leads of the winding can enter and exit the magnetic core. However, in the magnetic model, the geometry of the core is assumed to be perfectly axisymmetric. Moreover, the indentations cause the magnetizing and leakage inductances to be dependent on the position during the rotation of the secondary core. The influence of the indentations on the transferred power, iron losses, and the corresponding discrepancy with the 2D model are investigated using a 3D FEM model.

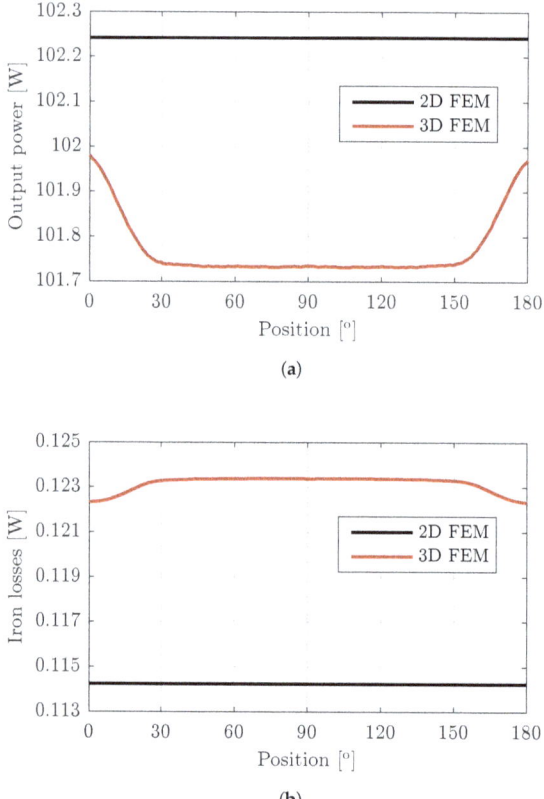

Figure 9. Comparison between the 2D and 3D model: (**a**) Output power as a function of position, and (**b**) iron losses as a function of position.

The results of the comparison are shown in Figure 9, where the output power and iron losses as a function of the position of the secondary core are shown in Figure 9a,b, respectively. At the zero position, the winding indentations are aligned, as shown in Figure 1. As a result of the rotation, the winding indentations misalign and the magnetizing inductance decreases, whereas the leakage inductances and the magnetic flux density in the remainder of the core geometry increase. Consequently, the output power is decreased by 0.24% and the iron losses increase by 0.86%. Therefore, the assumption of neglecting the influence of the rotation of the secondary core is justified. Furthermore, the average discrepancy between the 2D and 3D model for the output power and iron loss calculations are equal to +0.46% and −7.3%, respectively. Despite the relatively high discrepancy in the calculation of the iron losses, the effect on the efficiency and core temperature is negligible, since the amplitude of the iron losses is very small compared to the output power. Consequently, also the assumption of neglecting the winding indentations and modeling the geometry as an axisymmetric two-dimensional problem is justified.

4.2. Recommendations

Recommendations for future research include; firstly the improvement of the thermal model by the inclusion of heat transfer to the PCB, such that the estimation of the core temperature can be improved. Consequently, the core inertia could potentially be further reduced in a new geometrical optimization. Secondly, the design approach can be made more generic by the substitution of the 2D magnetic model for a 3D model, such that the accuracy of the core loss calculation is improved. Lastly, the system efficiency could potentially be improved by selecting more efficient diodes, since the largest portion of the losses occurs in the single-phase rectifier.

5. Conclusions

Both resonant and non-resonant high-frequency rotary transformers have been designed and compared within an electrical frequency range up to and including 1 MHz. The objective was to realize an output power level of at least 100 W. A multi-physical design approach has been proposed, in which magnetic, electrical, and thermal models are coupled. A design optimization for two fixed pot core geometries (P14/8 and P18/11 pot cores, respectively) in which the efficiency was maximized, has indicated that the application of series–series resonance within the investigated frequency range, increases the output power on average by 39.7% and 45.5%, respectively.

A geometrical optimization, in which the core inertia was minimized for the desired output power, has indicated that with respect to the optimal non-resonant design, the optimal resonant design reduces the core inertia by 45.5%, while the output power and efficiency are increased by 9.3% and 0.4%, respectively. Furthermore, with respect to the fixed core geometry, improvements in terms of core inertia of 38.2% and 66.4% are obtained by the non-resonant and resonant design, respectively.

The multi-physical design approach has been experimentally verified and closely matches the measurements; maximum discrepancies between the model and measurement results of +0.49%, −3.1%, +5.4% and +10.1% were obtained in the output power, efficiency, secondary and primary core temperature, respectively. Therefore, the multi-physical design approach has proven to be accurate and well-suited for the design of high-frequency WPT systems.

Author Contributions: The results presented in this paper have been developed by K.B. The analysis has been performed in cooperation with D.C.J.K., S.J., and E.A.L. The paper was written by K.B., and contributions and improvements to the content have been made by D.C.J.K., S.J., and E.A.L. All authors have read and agreed to the published version of the manuscript

Funding: This research received no external funding.

Acknowledgments: The author would like to thank Bram Daniels for his help in realizing multiple design optimization routines to be evaluated in parallel.

Conflicts of Interest: The authors declare no conflict of interest.

Abbreviations

The following abbreviations are used in this manuscript:

FEM	Finite Element Method
GaN	Gallium-Nitride
PCB	Printed Circuit Board
rms	Root-Mean-Square
WPT	Wireless Power Transfer

References

1. Campi, T.; Cruciani, S.; Maradei, F.; Feliziani, M. Innovative Design of Drone Landing Gear Used as a Receiving Coil in Wireless Charging Application. *Energies* **2019**, *12*, 3483. [CrossRef]
2. Zhang, C.; Lin, D.; Hui, S.Y.R. Ball-Joint Wireless Power Transfer Systems. *IEEE Trans. Power Electron.* **2018**, *33*, 65–72. [CrossRef]
3. Bao, Y.; Zhong, Z.; Hu, C.; Qin, Y. Rotor Field Oriented Control of Resonant Wireless Electrically Excited Synchronous Motor. *World Electr. Veh. J* **2019**, *10*, 62. [CrossRef]
4. Maier, M.; Parspour, N. Operation of an Electrical Excited Synchronous Machine by Contactless Energy Transfer to the Rotor. *IEEE Trans. Ind Appl.* **2018**, *54*, 3217–3225. [CrossRef]
5. Bortis, D.; Fässler, L.; Looser, A.; Kolar, J.W. Analysis of Rotary Transformer Concepts for High-Speed Applications. In Proceedings of the 28th Annual IEEE Applied Power Electronics Conference and Exposition (APEC), Long Beach, CA, USA, 17–21 March 2013; pp. 3262–3269.
6. Chen, X.Y.; Jin, J.X.; Zheng, L.H.; Wu, Z.H. A Rotary-Type Contactless Power Transfer System Using HTS Primary. *IEEE Trans. Appl. Supercond.* **2016**, *26*, 1–6. [CrossRef]
7. Qian, W.; Zhang, X.; Fu, Y.; Lu, J.; Bai, H. Applying Normally-Off GaN HEMTs for Coreless High-Frequency Wireless Chargers. *CES TEMS* **2017**, *1*, 418–427. [CrossRef]
8. Kazmierkowski, M.P.; Moradewicz, A.; Duarte, J.; Lomonova, E.; Sonntag, C. Contactless Energy Transfer. In *Power Electronics and Motor Drives*, 2nd ed.; Wilamowski, B., Irwin, J., Eds.; CRC Press: Boca Rotan, FL, USA, 2011; Chapter 35, pp. 1–17, ISBN 978-1-4398-0285-4.
9. Shin, J.; Shin, S.; Kim, Y.; Ahn, S.; Lee, S.; Jung, G.; Jeon, S.J.; Cho, D.H. Design and Implementation of Shaped Magnetic-Resonance-Based Wireless Power Transfer System for Roadway-Powered Moving Electric Vehicles. *IEEE Trans. Ind. Electron.* **2014**, *61*, 1179–1192. [CrossRef]
10. Trevisan, R.; Costanzo, A. A 1-kW Contactless Energy Transfer System Based on a Rotary Transformer for Sealing Rollers. *IEEE Trans. Ind. Electron.* **2014**, *61*, 6337–6345. [CrossRef]
11. Godbehere, J.; Hopkins, A.; Yuan, X. Design and Thermal Analysis of a Rotating Transformer. In Proceedings of the IEEE International Electric Machines and Drives Conference (IEMDC), San Diego, CA, USA, 12–15 May 2019; pp. 2144–2151.
12. Smeets, J.P.C.; Krop, D.C.J.; Jansen, J.W.; Hendrix, M.A.M.; Lomonova, E.A. Optimal Design of a Pot Core Rotating Transformer. In Proceedings of the IEEE Energy Conversion Congress and Exposition, Atlanta, GA, USA, 12–16 September 2010; pp. 4390–4397.
13. Ditze, S.; Endruschat, A.; Schriefer, T.; Rosskopf, A.; Heckel, T. Inductive Power Transfer System with a Rotary Transformer for Contactless Energy Transfer on Rotating Applications. In Proceedings of the IEEE International Symposium on Circuits and Systems (ISCAS), Montreal, QC, Canada, 22–25 May 2016; pp. 1622–1625.
14. Legranger, J.; Friedrich, G.; Vivier, S.; Mipo, J.C. Comparison of Two Optimal Rotary Transformer Designs for Highly Constrained Applications. In Proceedings of the IEEE International Electric Machines and Drives Conference (IEMDC), Antalya, Turkey, 3–5 May 2007; pp. 1546–1551.
15. Zhong, H.; Zhao, L.; Li, X. Design and Analysis of a Three-Phase Rotary Transformer for Doubly Fed Induction Generators. *IEEE Trans. Ind. Appl.* **2015**, *51*, 2791–2796. [CrossRef]

16. Zietsman, N.L.; Gule, N. Optimal Design Methodology of a Three Phase Rotary Transformer for Doubly Fed Induction Generator Application. In Proceedings of the IEEE International Electric Machines and Drives Conference (IEMDC), Coeur d'Alene, ID, USA, 10–13 May 2015; pp. 763–768.
17. Sudhoff, S. *Power Magnetic Devices*; John Wiley & Sons, Ltd.: Hoboken, NJ, USA, 2014; ISBN 978-1-118-48999-4.
18. Altair Engineering, Inc. Available online: https://www.altair.com/flux/ (accessed on 10 February 2020).
19. van den Bossche, A.; Valchev, V.C. *Inductors and Transformers for Power Electronics*; Taylor & Francis Group: New York, NY, USA, 2005; ISBN 978-1-4200-2728-0.
20. Fair-Rite Products Corp. Available online: https://www.fair-rite.com/78-material-data-sheet/ (accessed on 10 February 2020).
21. Bertotti, G. *Hysteresis in Magnetism*; Academic Press, Inc.: San Diego, CA. USA, 1998; ISBN 978-0-12-093270-2.
22. Efficient Power Conversion Corporation. Available online: http://epc-co.com/epc/Portals/0/epc/documents/guides/EPC9035_qsg.pdf (accessed on 10 February 2020).
23. Nexperia. Available online: https://www.nexperia.com/products/diodes/schottky-rectifiers/medium-power-low-vf-schottky-rectifiers-single-200-ma/PMEG6030ETP.html (accessed on 10 February 2020).
24. Smeets, J.P.C. Contactless Transfer of Energy: 3D Modeling and Design of a Position-Independent Inductive Coupling Integrated in a Planar Motor. Ph.D. Thesis, Eindhoven University of Technology, Eindhoven, The Netherlands, 19 March 2015.
25. Sinha, D.; Sadhu, P.K.; Pal, N.; Bandyopadhyay, A. Computation of Inductance and AC Resistance of a Twisted Litz-Wire for High Frequency Induction Cooker. In Proceedings of the International Conference on Industrial Electronics, Control and Robotics, Orissa, India, 27–29 December 2010; pp. 85–90.
26. Tang, X.; Sullivan, C.R. Optimization of Stranded-Wire Windings and Comparison with Litz Wire on the Basis of Cost and Loss. In Proceedings of the IEEE 35th Annual Power Electronics Specialists Conference, Aachen, Germany, 20–25 June 2004; pp. 854–860.
27. Pack Litz Wire. Available online: https://www.packlitzwire.com/products/litz-wires/rupalit-classic/ (accessed on 10 February 2020).
28. Analog Devices. Available online: https://www.analog.com/en/design-center/design-tools-and-calculators/ltspice-simulator.html (accessed on 10 February 2020).
29. Development Board EPC9035Quick Start Guide. Available online: https://epc-co.com/epc/Products/eGaNFETsandICs/EPC2022.aspx (accessed on 10 February 2020).
30. Holman, J. *Heat Transfer*; The McGraw-Hill Companies, Inc.: New York, NY, USA, 2010; ISBN 978-0-07-352936-3.
31. Wong, H.Y. *Handbook of Essential Formulae and Data on Heat Transfer for Engineers*; William Clowes & Sons Limited: London, UK, 1977; ISBN 0-582-46050-6.
32. Hendershot, J.R.; Miller, T.J.E. *Design of Brushless Permanent-Magnet Machines*; Motor Design Books LLC: Venice, FL, USA, 2010; ISBN 978-0-9840687-0-8.
33. Mathworks. Available online: https://nl.mathworks.com/help/optim/ug/constrained-nonlinear-optimization-algorithms.html#brnpd5f (accessed on 11 February 2020).

© 2020 by the authors. Licensee MDPI, Basel, Switzerland. This article is an open access article distributed under the terms and conditions of the Creative Commons Attribution (CC BY) license (http://creativecommons.org/licenses/by/4.0/).

MDPI
St. Alban-Anlage 66
4052 Basel
Switzerland
Tel. +41 61 683 77 34
Fax +41 61 302 89 18
www.mdpi.com

Energies Editorial Office
E-mail: energies@mdpi.com
www.mdpi.com/journal/energies

www.ingramcontent.com/pod-product-compliance
Lightning Source LLC
LaVergne TN
LVHW070609100526
838202LV00012B/606